Springer Theses

Recognizing Outstanding Ph.D. Research

Aims and Scope

The series "Springer Theses" brings together a selection of the very best Ph.D. theses from around the world and across the physical sciences. Nominated and endorsed by two recognized specialists, each published volume has been selected for its scientific excellence and the high impact of its contents for the pertinent field of research. For greater accessibility to non-specialists, the published versions include an extended introduction, as well as a foreword by the student's supervisor explaining the special relevance of the work for the field. As a whole, the series will provide a valuable resource both for newcomers to the research fields described, and for other scientists seeking detailed background information on special questions. Finally, it provides an accredited documentation of the valuable contributions made by today's younger generation of scientists.

Theses are accepted into the series by invited nomination only and must fulfill all of the following criteria

- They must be written in good English.
- The topic should fall within the confines of Chemistry, Physics, Earth Sciences, Engineering and related interdisciplinary fields such as Materials, Nanoscience, Chemical Engineering, Complex Systems and Biophysics.
- The work reported in the thesis must represent a significant scientific advance.
- If the thesis includes previously published material, permission to reproduce this must be gained from the respective copyright holder.
- They must have been examined and passed during the 12 months prior to nomination.
- Each thesis should include a foreword by the supervisor outlining the significance of its content.
- The theses should have a clearly defined structure including an introduction accessible to scientists not expert in that particular field.

More information about this series at http://www.springer.com/series/8790

Tek Prasad Adhikari

Photoionization Modelling as a Density Diagnostic of Line Emitting/Absorbing Regions in Active Galactic Nuclei

Doctoral Thesis accepted by
the Nicolaus Copernicus Astronomical Center
Polish Academy of Sciences (NCAC PAS),
Warsaw, Poland

 Springer

Author
Dr. Tek Prasad Adhikari
Nicolaus Copernicus Astronomical Center
Polish Academy of Sciences (NCAC PAS)
Warsaw, Poland

Supervisor
Prof. Agata Różańska
Nicolaus Copernicus Astronomical Center
Polish Academy of Sciences (NCAC PAS)
Warsaw, Poland

ISSN 2190-5053   ISSN 2190-5061   (electronic)
Springer Theses
ISBN 978-3-030-22739-5   ISBN 978-3-030-22737-1   (eBook)
https://doi.org/10.1007/978-3-030-22737-1

This Springer imprint is published by the registered company Springer Nature Switzerland AG
The registered company address is: Gewerbestrasse 11, 6330 Cham, Switzerland

Supervisor's Foreword

The interaction of radiation with matter is common phenomenon observed in the Universe in different energy bands. It is visible by detection of many absorption and emission lines from the variety of sources containing ionized gas. Dr. Tek Prasad Adhikari came to Poland from Kathmandu, Nepal, and started totally new topic of his research becoming a specialist on photoionization simulations of radiation in different active galactic nucleus (AGN) emitting/absorbing regions. The results of his thesis describe new, physically based methods to estimate the density of the astrophysical plasma surrounding the AGN. In addition, they address important questions on the nature of the physical processes underpinning two of the basic characteristics of the innermost regions around supermassive black hole: ionized outflows and emission line regions.

The main subject of his research is based on analysis of the warm absorber (WA) properties observed in X-rays and the nature of intermediate line regions (ILRs) recently detected in optical/UV band. In his work, Dr. Adhikari theoretically explained the nature and origin of absorption measure distribution (AMD) observed in AGN (currently, this property is observationally determined for eight sources). The subject is complex since for full analysis we need: from one side, high-resolution X-ray observations, and advanced data modelling, and from the second side, we have to make numerical simulations of photoionized gas in those sources. Dr. Adhikari achieved results in the second part of this topic. By making advanced numerical modelling, he found that the observed deeps in AMD are well explained by thermal instabilities in ionized gas. It appears to be the general property of these type of sources.

Dr. Adhikari also worked on the topic connected with AGN line emission regions. He managed to explain the possibility of the existence of intermediate line region (ILR) located between broad line region (BLR) and narrow line region (NLR) in AGN. This work was also done with the use of photoionization calculations made by Dr. Adhikari with high fluency. He found that by increasing the gas density of radially distributed clouds, observed gap of line emission between BLR and NLR is naturally reduced and ILR can be visible, which is in agreement with recent observations of several sources. The result does not depend much on AGN type, which was also clearly demonstrated by Dr. Adhikari.

His Ph.D. thesis is a collection of important findings showing that density matters for photoionization calculations, which was not believed by many people in X-ray community. This innovative work combines theoretical work on the inter-action of radiation with gas visible in at least two energy bands: X-rays and optical/UV in two different AGN regions. Interestingly, Dr. Tek Prasad Adhikari achieved the same consistent conclusions about the gas properties for those two widely observed AGN gas structures. It was a fantastic summary of his entire work made during his Ph.D. studies at the Nicolaus Copernicus Astronomical Center of the Polish Academy of Sciences (NCAC PAS) in Poland, and it only reminds me to express my satisfaction that I could be his scientific supervisor.

Warsaw, Poland   Prof. Agata Różańska
June 2019

Preface

This thesis is an outcome of my Ph.D. research work done at the Nicolaus Copernicus Astronomical Center of the Polish Academy of Sciences (NCAC PAN), Warsaw, Poland, during 2013 to 2018. My effort has been to throw light on the photoionization process that operates in the environment of the active galactic nuclei (AGNs). Particular focus has been given on the use of the existing photoionization codes Titan and Cloudy in the context of the absorbing and emitting medium in AGNs.

From the study of correlation between the continuum and lines variability in active galactic nuclei (AGNs), it is widely accepted that the photoionization is the principal mechanism of emission and absorption lines production. This effect is observed in various AGN components: broad line region, narrow line region, warm absorber, and dusty torus, where it is strongly believed that gas is photoionized by the AGN continuum radiation. One of the most important physical parameters is the location of the absorbing/emitting clouds from the SMBH, which is related to the gas density. Hence, the knowledge of gas density directly gives the location of the photoionized gas which is very essential to understand the AGN environment. The density of the emitting/absorbing plasma in AGNs is constrained from the various methods each of them having their own limitations. So, any new independent method of estimating the gas density in these regions is of particular importance. In this thesis, I show that the photoionization modelling of the ionized gas can be potentially used as a density diagnostic of the absorbing and emitting plasma in AGNs.

A brief overview of the AGN properties and a short review of the relevant literature is given in Chap. 1. Chapter 2 contains the description of the procedure commonly adopted in the calculations of the radiation transfer through the gas clouds. Particularly, I describe the parameters and the concepts used in the photoionization simulations done with Cloudy and Titan. The density dependence of the photoionization models and its consequences in the modelling of the warm absorption properties and the line-emitting properties are discussed in Chap. 3. The observational property of AGN and the absorption measure distributions (AMDs) is compared with that obtained from simulations done with the photoionization code

Titan in Chap. 4. Chapter 5 contains a study of the newly observed intermediate line region (ILR) using the numerical code CLOUDY. I demonstrate that both the AMDs and the line emissivity radial profiles depend on the gas density which opens a new potential method of density diagnostic. Additionally, I show that the newly observed intermediate line region (ILR) can be successfully recovered from the photoionization simulations by assuming the high gas density of this region. A summary and the possible future extensions of this work are presented in Chap. 6.

Successful completion of this thesis is triggered by various people who assisted me throughout my studies at the Nicolaus Copernicus Astronomical Center. First and the foremost, I would like express very great appreciation to my supervisor Dr. hab. Agata Różańska for the fruitful scientific discussion and the constant encouragement. Discussion with her has always generated a new insight into the subject matter. Without her scientific vision, continuous encouragement, and correct approach to understand the subject matter of the present research, this thesis would never take the present shape.

I am thankful to Prof. Bozena Czerny for her scientific comments and suggestions, which played a vital role in the completion of the projects presented in this thesis. Special thanks go to my collaborator Dr. Krzysztof Hryniewicz for his scientific suggestions and especially for helping to solve the problems in Python. I would like to thank Prof. Gary Ferland for providing many crucial suggestions regarding the code Cloudy. I am thankful to Prof. Anne-Marie Dumont for her helpful suggestions for the efficient use of the Titan code. I would like to thank my friend Swayamtrupta Panda for his suggestions regarding the English correction. The constructive comments of my thesis review committee members: Dr. Matteo Guainazzi and Dr. Andrzej Niedźwiecki have played a vital role for improving the quality of this work.

Special acknowledgement goes to the Scientific Council of the NCAC PAN for nominating my thesis for the prestigious Springer Thesis Award. Thank you Springer for giving this honor. Thanks to all the professors, the scientific staffs, and my fellow Ph.D. students of the Nicolaus Copernicus Astronomical Center with whom I got an opportunity to understand many different scientific problems via regular discussions, lectures, and seminars. I would like to acknowledge all the administrative staff, library staff, and computer department staff for their friendly behavior and constant help throughout my studies.

Last but not least, I would like to remember my family members for their precious support during my studies. In particular, my lovely wife Kabita and my parents have suffered a lot by permitting me to live six thousands kilometers away and continue my studies. Thanks for all your support.

Kathmandu, Nepal   Dr. Tek Prasad Adhikari
March 2019

Publications

Most of the results contained in this thesis are presented in the following publications:

Refereed Publications

1. **Adhikari T. P.**; Różańska A.; Sobolewska M.; Czerny B. *Absorption Measure Distribution in Mrk 509*, 2015, ApJ, 815, 83A, (Chapter 4)

2. **Adhikari T. P.**; Różańska A.; Czerny B.; Hryniewicz K.; Ferland G. J. *The Intermediate Line Region in AGN*, 2016, ApJ, 831, 68A, (Chapter 5)

3. **Adhikari T. P.**; Różańska A.; Hryniewicz K.; Czerny B.; Ferland G. J. *On the Intermediate Line Region in AGNs*, 2017, Frontiers in Astronomy and Space Sciences 4, 19, (Chapter 5)

4. **Adhikari T. P.**; Hryniewicz K.; Różańska A.; Czerny B.; Ferland G. J. *Intermediate line emission in AGN: the effect of gas density prescription*, 2018, ApJ, 856, 78A (Chapter 5)

Conference Proceedings

1. **Adhikari T. P.**; Różańska A.; Sobolewska M.; Czerny B. *On the warm absorber in AGN outflow*, 2016, Proceedings of the XXXVII Polish Astronomical Society Meeting, 3, 239, (Chapter 4)

2. **Adhikari T. P.**; Różańska A.; Hryniewicz K.; Czerny B. *Absorption Measure Distribution in AGN*, 2017, accepted for publication in the Proceedings of the XXXVIII Polish Astronomical Society Meeting, in press, (Chapter 4)

Contents

Abbreviations

AAS	American Astronomical Society
AD	Accretion disk
AGN	Active galactic nucleus
AMD	Absorption measure distribution
BAL	Broad absorption line
BBB	Big blue bump
BLR	Broad line region
BLLac	BL Lacertae
CD	Constant density
COS	Cosmic Origins Spectrograph
CP	Constant pressure
EPIC	European Photon Imaging Camera
FR-I	Fanaroff-Riley class I
FR-II	Fanaroff-Riley class II
FWHM	Full width of half maximum
HETGS	High Energy Transmission Grating Spectrometer
HIL	High-ionization lines
HST	Hubble Space Telescope
ILR	Intermediate line region
IRAS	Infrared Astronomical Satellite
ISM	Interstellar medium
LIL	Low-ionization lines
LINER	Low-ionization nuclear emission region
LLAGN	Low-luminosity active galactic nucleus
LTE	Local thermodynamic equilibrium
MOS	Metal oxide silicon (MOS)
NLR	Narrow line region
NLSy1	Narrow-line Seyfert 1 galaxy
OVV	Optically violent variable
OM	Optical monitor

RGS	Reflection Grating Spectrometer
RM	Reverberation mapping
RPC	Radiation Pressure Confined
Sy1	Seyfert 1 galaxy
Sy2	Seyfert 2 galaxy
SED	Spectral energy distribution
SMBH	Supermassive black hole
QSO	Quasi-stellar object
WA	Warm absorber
UFO	Ultra-fast outflow
UV	Ultra violet
UVOT	Ultraviolet/optical telescope
XRT	X-ray telescope

Physical Constants and Symbols

Constants Used

c Speed of light (2.997×10^{10} cm s^{-1})
k Boltzmann constant (1.3807×10^{-16} cm^2 g s^{-2} K^{-1})
$M_\odot$ Mass of the Sun (1.989×10^{33} g)
σ_T Thomson scattering cross section (6.652×10^{-25} cm^2)
G Universal gravitational constant (6.674×10^{-8} cm^3 g^{-1} s^{-2})
h Planck's constant (6.626×10^{-27} erg s)

Symbols Used

A Density normalization (cm^{-3})
$A_{Z_\odot}$ Solar abundances (dimensionless)
α_{OX} Soft excess parameter (dimensionless)
α_B Recombination coefficient (cm^{-3} s^{-1})
α_G Recombination coefficient of ground level (cm^{-3} s^{-1})
Γ Power law photon index (dimensionless)
$f(E)$ Band function flux (erg s^{-1} cm^{-2})
E_p Peak energy of the Band function (eV)
α Slope of first power law of Band function (dimensionless)
β Slope of the density power law (dimensionless)
γ Slope of second power law of Band function (dimensionless)
r Distance from SMBH (cm)
R_d Dust sublimation radius (cm)
R_{BLR} Broad line region radius (cm)

P_{gas} Gas pressure (dyn cm^{-2})

P_{rad} Radiation pressure (dyn cm^{-2})

ξ Ionization parameter (erg cm s^{-1})

Ξ Dynamical ionization parameter (dimensionless)

U Dimensionless ionization parameter (dimensionless)

I_ν Specific intensity (erg s^{-1} cm^{-2} sr^{-1} Hz^{-1})

S_ν Source function (erg s^{-1} cm^{-2} sr^{-1} Hz^{-1})

j_ν Emission coefficient (erg s^{-1} cm^{-3} sr^{-1} Hz^{-1})

J_ν Mean intensity of radiation field (erg s^{-1} cm^{-2} Hz^{-1})

g Acceleration due to gravity (cm s^{-2})

$L_\odot$ Luminosity of the Sun (erg s^{-1})

L_{ion} Ionizing luminosity (erg s^{-1})

L_{disk} Disk luminosity (erg s^{-1})

L_X X-ray luminosity (erg s^{-1})

n_H Hydrogen gas density (cm^{-3})

n_e Electron density (cm^{-3})

n_p Proton density (cm^{-3})

ϕ_H Flux of H-ionizing photons (cm^{-2} s^{-1})

μ Cosine of zenithal angle (dimensionless)

ρ Volume gas density (g cm^{-3})

κ Mean opacity coefficient (cm^2 g^{-1})

κ_ν Absorption coefficient (cm^2 g^{-1})

σ_ν Scattering coefficient (cm^2 g^{-1})

F_{ion} Ionizing flux (erg s^{-1} s^{-2})

f_{ion} Fractional ion abundances (dimensionless)

Γ_{tot} Total heating rate (erg cm^3 s^{-1})

Γ_C Compton heating rate (erg cm^3 s^{-1})

Λ_{tot} Total cooling rate (erg cm^3 s^{-1})

Λ_C Compton cooling rate (erg cm^3 s^{-1})

Λ_{FF} Free-free cooling rate (erg cm^3 s^{-1})

z Geometrical depth (cm)

τ Optical depth (dimensionless)

ρ Gas volume density (g cm^{-3})

N_H Hydrogen column density (cm^{-2})

T_{IC} Compton temperature (K)

T Temperature (K)

l Thickness of the ionization front (cm)

M_{bh} Mass of the black hole ($M_\odot$)

Q_H Number of H-ionizing photons (s^{-1})

$\dot{m}$ Eddington ratio (dimensionless)

$H\alpha$ Hydrogen Lyman α line

σ_{tot} Total scattering cross section (cm^2)

g_{ff} Mean Gaunt factor (dimensionless)

Chapter 1
Introduction

Abstract The best way to study the supermassive black holes (SMBHs) is to investigate the matter in the AGN environment. The radiation, emitted as a result of the accretion, interacts with and ionizes the surrounding materials, the spectral signatures of which can be seen in the observed optical/UV/X-ray spectra of AGNs. With the advent of observational techniques from radio to the hard X-ray bands, the electromagnetic properties of AGNs have been known in much details over the period of a few decades. One of the most important achievements from multi-wavelength observations of AGNs is the identification of various components: the accretion disk (AD) surrounding the SMBH, the emission regions where the broad and narrow lines originate, the absorption regions, the collimated and dispersed outflows, the dusty torus, and the hot corona where hard X-ray emission comes from. This thesis deals with the two major components; the absorption regions and the emission regions. A brief overview of the AGN properties and a short review of the relevant literature is given in this Chapter.

Active galactic nuclei (AGNs) are the bright and luminous center of the galaxies believed to harbor supermassive black holes (SMBHs) of a few million to few billion of Solar masses. Whereas it is likely that all galaxies contain SMBH at their centers, the AGNs are spectacular in a way that high amount of persistent radiation energy 10^{12} to 10^{14} $L_\odot$ is emitted from them. The most likely scenario for the production of such huge luminosity is the conversion of gravitational energy of matter accreted by the black hole to the mechanical and electromagnetic energy [1]. The most efficient model of accretion is through the circular, geometrically thin, optically thick disk [2] formed around the black hole, where the angular momentum is carried outwards as a result of the flow of matter inwards. All other known energy production mechanisms are inefficient to explain the observed luminosities of the AGNs. It is estimated that 10–30% of total energy radiated in the Universe is generated due to accretion process [3].

The best way to study SMBHs is to investigate the matter in the AGN environment. The radiation, emitted as a result of the accretion, interacts with an ionizes the surrounding materials, the spectral signatures of which can be seen in the observed optical/UV/X-ray spectra of AGNs. With the advent of observational techniques from

© Springer Nature Switzerland AG 2019

T. P. Adhikari, *Photoionization Modelling as a Density Diagnostic*
of Line Emitting/Absorbing Regions in Active Galactic Nuclei,
Springer Theses, https://doi.org/10.1007/978-3-030-22737-1_1

radio to the hard X-ray bands, the electromagnetic properties of AGNs have been known in much details over the period of a few decades. One of the most important achievements from multi-wavelength observations of AGNs is the identification of various components: the accretion disk (AD) surrounding the SMBH, the emission regions where the broad and narrow lines originate, the absorption regions, the collimated and dispersed outflows, the dusty torus, and the hot corona where hard X-ray emission comes from.

The observational study of AGNs in the nearby Universe is quite successful in deriving the different components defining this system. The radiative energy continuum from AGN spans the broad electromagnetic spectrum from radio band upto the γ-rays. From the analysis of different parts of the observed spectrum, it has been well established that a large part of the emission is produced in the AD and it is emitted in the optical and UV bands. A large fraction of these seed photons produced in the AD are reprocessed in the dusty torus located beyond the dust sublimation radius and re-emitted in the infrared band. Also the hot electrons in the region close to the AD scatters the fraction of photons coming from the AD and gives rise to X-rays. So, by studying the broad spectral energy distribution of AGNs, one can probe various components forming the whole complex AGN system.

The broad band AGN continuum exhibits numerous absorption and emission lines which are detected in the spectrum. These observed emission and absorption lines are most likely produced from the interaction of continuum radiation with the surrounding materials i.e. by the process of photoionization. The direct evidence to support this idea of line production is the correlation between the line and the continuum variations in many AGNs, where the continuum luminosity changes are followed by the line luminosity changes.

The state-of-art numerical simulations are very essential to incorporate the existing physical models for mimicking the processes that are responsible for the observed properties of AGNs. This thesis deals with the photoionization process which is generally accepted to be the main mechanism to produce the emission and absorption lines in the observed AGN continuum. By a careful comparison of the results of the numerical simulations to that of observations, I show that the main physical properties: the ionization structure and the gas density of the matter in the vicinity of the SMBH can be satisfactorily constrained.

1.1   Observational Properties of AGNs

The observational studies of AGNs date back to the work of Edward A. Fath in 1908 in the Lick observatory [4], where he was studying the spectra of spiral nebulae, now known to be galaxies, using the photographic slitless spectrograph. He found six emission lines in the spectrum of NGC 1068, one of them Hβ known at that time and others as known today are [O II] $\lambda3727$, [Ne III] $\lambda3869$, and [O III] $\lambda4363$, 4959, 5007. It was only after 1926, Edwin Hubble resolved the issue of famous 'Shapley–Curtis Debate' of 1920 showing that the spiral nebulae are of extragalactic nature [5]. Later, Seyfert [6] published a paper discovering more galaxies, for example NGC 1068, and showed that a small fraction of galaxies display many high ionization emission lines. These galaxies with broad emission lines were later recognized as Seyfert galaxies. Since then, tremendous efforts have been devoted in the observation of AGNs using multi-wavelength instruments.

The broadness of the emission lines is measured in terms of the full width at half-maximum (FWHM). The pioneering optical observations of AGN made by Carl Seyfert, showed that some sources exhibit broad lines with FWHM $> 2000 \, \text{km s}^{-1}$ in their spectra. These broad lines are produced in a dust free region at subparsec scales from the central SMBH named broad-line region (BLR). The narrow lines with FWHM $< 500 \, \text{km s}^{-1}$ that are ubiquitous among the spectra of AGNs are produced in the narrow-line region (NLR) which extends from several pc to few kpc scale. Below, I list the general classification of AGNs based on the differences in their observational properties in radio/optical/UV domain, while in the following subsections, I present more detailed characterization of most important AGN components studied in this thesis:

- Radio quiet AGNs: This class of AGNs show very weak/no radio emission. Depending on other spectral properties they are further classified into several types such as: Seyfert galaxies and low ionization nuclear emission regions (LINERs). Seyfert galaxies are further subdivided into Seyfert 1 (Sy1)/Seyfert 2 (Sy2) types depending on the presence/absence of the broad emission lines. Narrow emission lines are prevalent among the spectra of both types. Sy1 galaxies with narrower broad lines are also called narrow line Sy1 (NLSy1). LINERs show only the weak nuclear emission line regions and show very low activity of accretion. It is not yet resolved whether all LINERs are AGNs or not [7]. The more distant and more luminous objects with properties similar to Seyfert galaxies form another class of AGNs called quasars (QSOs).
- Radio loud AGNs: This type of AGNs exhibit strong radio emission originated in the well collimated bipolar jets of the material moving away from the AGN center. Other spectral properties are very similar to that of the radio quiet AGNs. The objects with the lack of optical/UV disk component in their spectra are called BL Lac objects. Such objects possess strong jet directed towards the observer. In some cases the optical emission can be visible when the emission is variable and such objects are called optically violent variable (OVV) quasars. The OVV and BL Lac objects are also commonly known as blazars. Among this class, the objects having radio emission mostly from the core of the jet are called Fanaroff-Riley Class I (FR-I) and those where the radio emission comes from the part of the jet further away are called Fanaroff-Riley Class II (FR-II) types [8].

In this thesis, I consider only radio quiet AGNs. Since, I study emitting/absorbing regions located within few tens of parsecs from SMBH, I do not take illumination of those regions by radio emission, which originates from much larger regions.

1.1.1   Unified Theory of AGN

The unified theory [9, 10] basically tells that all AGN types belong to the same parent population of the AGN with similar inherent properties. The difference in the observed properties is due to the different inclination angles at which they are observed. The inclination angle here is defined as the angle between the line of sight and the rotation axis of the AD. The development of the unified theory was motivated by the discovery of highly polarized broad component of Hα line in the galaxy 3C 234 by Antonucci [11]. The observed polarized spectra can be successfully explained if the central engine is obscured by a dusty toroidal structure commonly known as the dusty torus. The torus scatters the direct emission from the BLR and is responsible for the observed polarized spectra. Based on the observational properties of different types of AGNs, the different components of AGN unified model can be summarized as:

- Accretion disk (AD): Rapidly rotating gas in the gravitational field of the SMBH forms an AD through which the matter is slowly moving onto the black hole. Accretion disk is prevalent among the AGNs, proto-planetary disks, stellar binaries where the gravitational energy is constantly being converted into the heat, a fraction of which is converted into the radiation. In AGNs, most of the seed photons are produced in the AD and radiated in optical/UV band. This is displayed by the big hump also known as big blue bump (BBB) in the broad band spectral energy distribution (SED) of AGNs. Although there exists number of proposed theoretical models of the accretion disks, the standard model that explains observational properties of the disk is the optically thick and geometrically thin disk model proposed by Shakura and Sunyaev [2].
- Broad line region (BLR): The broad emission lines present in the optical and UV spectra are the basic signatures of active galaxies and an important probe of how the AGN system works. The major broad emission lines identified in the AGN spectra are: the permitted transitions of Mg II, Hα, Hβ, Fe II with their FWHM ≥ 2000 km s^{-1}. The observational studies over the past 50 years have shown that the gas which gives rise to the broad line emission is located close to the AD and it is predominantly in the Keplerian motion with additional velocity components [12, 13]. The additional contribution in the motion is often described due to the possible spiral structures in the disk as well as strong turbulent motions, inflows and outflows. BLR is very useful tool to determine the mass of the black hole as well as for the study of gas dynamics in its neighbourhood. The dust probably cannot survive in BLR due to the intense radiation from the central part of AD.
- Narrow line region (NLR): In addition to the presence of broad emission lines in the AGN spectra, narrow emission components with FWHM ~ 500 km s^{-1} are also commonly seen in the observations. The region where these narrow lines are produced is called the narrow line region which is located at much further distance from the SMBH. And therefore the radiation from the disk is not strong enough to evaporate the dust in this region.

- Torus: The first sketch of the circumnuclear toroidal structure of the obscuring gas and dust was given by Antonucci [11] forming a cornerstone of the AGN unification scheme. As this structure is optically thick, it is responsible for obscuring the direct emission from the BLR in Sy2 galaxies. The polarized emission observed in some AGNs can be well explained by the effect when the central engine is obscured by a dusty torus. As constrained from Infrared imaging [14], interferometry [15] and very recently by using sub-millimeter observations [16, 17], the outer boundary of the torus lies at the distance range 0.1–10 pc. It is also responsible for the collimation of the radiation and hence producing the biconical shapes of the NLR and of the ionization cones [18].
- Outflows: The outflows in AGNs provide a convincing way of connecting the accretion energy of the SMBH to the interstellar medium of the host galaxy. By knowing the properties of the outflows, one can ponder the feedback that is provided to the physical processes occurring in the host galaxy. Outflows in AGN occur in different flavours. One of the most prominent types is the collimated bipolar outflows also known as jets which are prevalent in radio loud galaxies. The strong radio emission in such galaxies is produced due to the Synchrotron process in which the acceleration of charged particles by magnetic field is the reason of the emission. The collimated outflows are present in radio loud AGNs where most of the radio power is emitted in jets [19–21].

Since unified model has been established, new observations provided evidences for new spectral components in those objects. Below, I list components which are well known in many AGN, but they do not give yet any constraints on the source inclination i.e. type of AGN according to the unified model.

- X-ray Corona: For many years it is known that the strong X-ray emission is ubiquitous in AGNs and is believed to be produced in the compact and hot region located within a few gravitational radii from an AD above the black hole [22]. The seed photons from an AD are scattered and reprocessed in the corona producing X-rays. The most possible explanation of this process is the Comptonization which produces broad band X-ray continuum with cut off at higher energies upto several hundreds of keV, limited by the Klein-Nishina process in relativistic electrons. The observed X-ray spectrum of AGNs is characterized by the power law of average photon index $\Gamma = 1.9$ [23], with typical variation over $\Gamma = 1.7$–2.0 in the range 0.5–10 keV. This is accompanied by a broad hump seen over 20–40 keV above the canonical power-law and this is generally described as due to the reflection of hard X-ray photons, produced in the corona, from the relatively colder AD [24]. The recent hard X-ray and gamma-ray telescopes have discovered an exponential cut-off around the energy as high as few hundred keV to the power-law extrapolated from lower energies. This is mostly due to the fact that at these high energies the energy gain in each Compton scattering is less than the energy lost by the photons during the electron recoil.
- Ionized outflows: Uncolimated outflows in AGNs are detectable due to absorption features on ionized heavy elements resolved in high resolution optical/UV and X-ray observations of those objects. Over 20% of quasars exhibit broad absorption

lines (BALs) in their optical/UV spectra. All BALs can exhibit very complex blueshifted profiles, indicating the presence of several absorbing systems moving toward the observer with different velocities, as is in QSO 2359-1241 for example [25]. Narrow line components usually indicate velocities of the order of a few hundred km s^{-1}, while velocities of some broad lines can reach even 50,000 km s^{-1} [26]. Similar UV absorption is observed in Sy galaxies [27]. Furthermore, in about 50% of those objects uncollimated outflows are inferred from X-ray observations [28–32], where the X-ray spectra exhibit numerous absorption lines which are systematically outflowing with velocities in the range 10^2 to 10^3 km s^{-1}. Those lines originate from highly ionized gas named commonly as warm absorber (WA). More recently, highly blueshifted Fe K-shell absorption lines at energy ≥ 7 keV have been detected in the X-ray spectra of several AGNs [33–38]. This class of outflow is referred as the Ultra Fast Outflow (UFO), since estimated absorbing gas velocity is set to be up to 0.3 of the velocity of light. There exist debating arguments whether UFOs and WAs are the same outflows observed at the different distances from the SMBH [36], or they are of inherently different origins [37].

- Intermediate line region (ILR): In the recent observations of some AGNs, intermediate line component of FWHM ~ 700–1200 km s^{-1}, in addition to the broad and narrow line component, is clearly required to fit the emission lines in their spectra [39–45]. These findings changes the traditional understanding of the radial emissivity properties of the AGNs where no emission is seen from a region between BLR and the NLR. The open questions are: does the ILR exist physically separated from BLR and NLR? What are the mechanisms that give rise to ILR in some sources but not in others?

In this thesis, I will focus my study on the WA flavour of the outflow observed with high resolutions X-ray missions. Furthermore, I will provide evidences of existence of the ILR region located between BLR and NLR of some AGNs.

1.1.2 Warm Absorbing Gas

The evidence for the X-ray absorption of the continuum radiation in AGN was first shown by Halpern [46] in the *EINSTEIN* data of QSO MR 2251-178. This conclusion was derived based on O VII and O VIII absorption edges in the observed X-ray spectrum. The most feasible explanation for the absorption was the presence of intrinsic warm gas located on the line of sight to the center of nucleus. It was only after the availability of the high resolution X-ray data, a significant progress has been made in the research of these ionized absorber. Now, there is a general consensus based on high resolution X-ray data that the majority (at least 50%) of the Seyfert galaxies contain ionized absorbing gas in their line of sight (Fig. 1.1).

Past fifteen years have witnessed the advancements in the X-ray astronomy as the results of high resolution observational data obtained using the space satellites *Chandra*, *XMM-Newton* and *Suzaku*. By using the gratings on board of first two

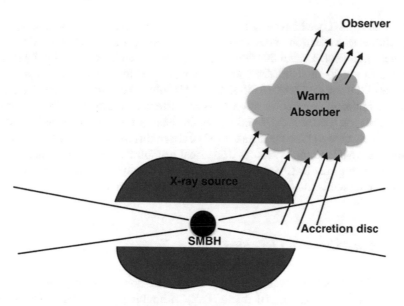

Fig. 1.1 Schematic diagram showing the warm absorber illuminated by radiation from the nucleus. An observer sees absorption lines in the spectrum transmitted through the absorber. The figure is reproduced from Adhikari et al. [62] with the reprint permission from the Polish Astronomical Society

satellites, many narrow absorption lines from highly ionized elements have been detected providing a great opportunity to study the WA in the vicinity of the SMBH [28–32, 36, 47–57].

From the measurements of energy shifts of the absorption line centroids, it has been established that the absorbing matter is systematically outflowing with velocities in the range 10^2 to 10^3 km s^{-1} [28, 48]. Directly from observations, ionic column densities of particular ions could be found using photoionization simulations, and they are typically of the order of 10^{15-18} cm^{-2}. Since, for a given source, we observe lines from wide range of ionization states, in practice ionic column densities are derived for each lines separately. For this purpose, photoionization calculations are done through thin, constant density (CD) slab of matter, assuming appropriate value of ionization parameter ξ (see Eq. 2.9 for definition).

In the next step, by assuming separate ionization zone for each ion, it is possible to calculate the equivalent hydrogen column densities. Such equivalent hydrogen column densities are in the range of 10^{18-23} cm^{-2} [32, 54]. Moreover, those column densities correspond to the continuous change of the ionization parameter which spans the range of few decades, when determined from the data of individual source.

Since then, several attempts have been made to model continuous ionization structure of the WA in several AGN [58–60]. Furthermore, Holczer et al. [58] proposed to describe ionization structure of the wind by determining the absorption measure distribution (AMD), which describes how hydrogen column density of ionizing material behaves as a function of the ionization degree along the line of sight (Eq. 2.13 in this thesis). In Sect. 2.3.2 of this thesis, I present detailed way of obtaining AMD from high resolution X-ray data. The above authors, for the first time, have shown that in the case of the source IRAS 13349+2438, AMD obtained from observations display deep minimum in column density that is consistent with the negligible absorption of gas with $\log \xi$ between 0.8 and 1.7. Such deep minima are present in AMD of other objects as well (see for instance [59–61]). In this thesis, I show that such deep minima in AMD are the evidences of thermal instabilities in a given ionization and temperature regime. In Chap. 4, I demonstrate what physical parameters affect the position of unstable zones and overall AMD normalization.

1.1.3 Line Emission Regions

The traditional picture, in the context of line emission in AGNs, consists of two physically separate regions: BLR and NLR, without any significant emission from the region between them. The non-existence of dust in the BLR implies that this region is much closer to the SMBH. Moreover from the reverberation mapping (RM) studies, it has been shown that the BLR radius, R_{BLR} scales with the AGN luminosity as $R_{BLR} \sim 0.01(L/10^{44})^{0.62-0.70}$ pc [63, 64]. On the other hand, it is well established that the dust is present in the NLR implying that it is located much further away from the SMBH. Using the variability studies of the forbidden line [O III] in NGC 5548, Peterson et al. [65] estimated the size of the NLR to be $\sim$2–3 pc.

The leading theoretical explanation for the suppression of emission in the transition region between BLR and NLR was given by Netzer and Laor [66] with an introduction of dust extinction, both in scattering and absorption of the line and continuum photons. The authors calculated the line emission from radially distributed clouds above an AD, using photoionization computations. Each cloud was a constant density slab illuminated by the same mean quasar continuum shape. The lack of line emission was successfully achieved with the introduction of dust. Practically it means, that the dust was taken into account in photoionization calculations for clouds located further away from SMBH at a certain radius named sublimation radius. Closer to the nucleus the radiation field is so strong that the dust grains cannot survive. The presence of dust for a given gas conditions successfully suppresses line emission, and the gap between BLR and NLR is naturally formed.

However, recent observations of several AGNs require the presence of ILR to fully explain observed line profiles. The statistical investigations of broad UV-lines in luminous QSOs show that the broad line region can be decomposed into a very broad component of FWHM $\geq 7000\,\mathrm{km\,s^{-1}}$ and a narrower component of width $\sim$2000 km s^{-1} [39]. The authors described that the narrower component arises in a distinct intermediate line region (ILR) that is an inner extension of the NLR. Later, in NLSy1 RE J1034 + 396, Mason et al. [40] have shown the evidence for the existence of significant amount of both permitted and forbidden emission line fluxes of

FWHM ~ 1000 km s^{-1}. For the Sy1 NGC 4151, Crenshaw and Kraemer [41] identified a line emission component with width FWHM $= 1170$ km s^{-1}, most probably originating between the BLR and NLR. Detailed spectral analysis of large number of *SDSS* sources have revealed the presence of intermediate component of line emission with velocity width in between that of broad and narrow components [42–44]. Additionally, Crenshaw et al. [67] detected ILR emission with width of 680 km s^{-1} in the historically low state spectrum of NGC 5548 obtained with the Space Telescope Imaging Spectrograph on the *Hubble Space Telescope*. Using the *HST/ FOS* spectra of quasar OI 287, Li et al. [45] reported the detection of intermediate width emission lines originating at a distance of ~ 2.9 pc from the central black hole. The question remains unanswered, is there any separation between gas responsible for broad and narrow line emission in different types of AGN? What is the physical mechanism leading to the existence of this gap in some source, and why the gap in line emission is not observed in all of them? This thesis addresses some aspects of these questions and I discuss my findings on the topic in the Chap. 5.

1.2 Density Diagnostic

The density of the ionized plasma is an important physical parameter in the study of the environment of the AGNs. The knowledge of the gas density directly gives the information about its location from the central engine of the AGNs. This is possible since the product of the density times the distance squared can be obtained from observations when the source luminosity is known, and when the ionization parameter is constrained from the line fitting procedure, (see Eq. 2.9 for exact definition). Nevertheless, the density times the distance squared product cannot be easily separated. The same value of ionization parameter is obtained for dense clouds located closer to the nucleus and the rare clouds located further away. Therefore, independent gas density measurements are needed to find the wind location. There exist several methods to estimate the density of the ionized gas in the AGN environment using observational as well as modelling techniques. Nevertheless each methods have their own limitations and assumptions which present further uncertainties in the distance estimation of the ionized gas.

One of the frequently implemented method of density estimation in case of WA is using the variability studies of the AGNs. The ionizing continuum of AGN can vary by more than a order of magnitude, on timescales as short as few hours. The effect induced by the source's variability on the behaviour of the gas is often described by the time dependent ionization balance equations, introduced by Krolik and Kriss [68]. The solution of such equation gives the response time, i.e. the time responsible for the gas to reach equilibrium with the ionizing continuum. In case of WA, Nicastro et al. [69] showed that such response timescale is inversely proportional to the density of gas. Thus studying the variability of WA yields an estimate of the density of the gas [69–71] which in turn allows for the estimation of its distance from the central source through Eq. 2.9. Nevertheless, while applying this method to estimate the density, two important simplifications: (i) all variations in gas properties are caused by the continuum variations and (ii) the shape of the continuum remains unchanged, are made.

The next important method used to estimate the gas density is by studying the density-sensitive absorption lines from the metastable levels. This method requires high quality UV spectra where the gas density can be estimated by comparing the ionic column densities of the excited metastable states of the most abundant ions HeI, CII, SiII, and FeII [72–74]. For the AGNs that were analyzed in the above papers the estimated number density is in the order of 10^3 to $10^{4.4}$ cm^{-3}. These values of densities put the gas location approximately in the range 3–6 kpc. However this method is very hard to apply for the X-ray absorber [75].

The gas density of the line emitting gas is usually estimated by studying the forbidden and permitted emission lines in the observed spectra of AGNs. The atomic transition probability for the permitted lines is very high as compared to the probability of transitions for the forbidden lines. If the gas density is less than the critical density (which depends on the transition probability), collisions do not de-excite the atoms and there is forbidden line emission. The presence of strong forbidden lines [O III] $\lambda 5007$Å from the narrow line region indicates that the gas density is less than $\sim 10^8$ cm^{-3}, the critical density for the [O III]. The estimation of the BLR density is more complex. On the one hand, the absence of the [O III] in the BLR indicates that the density is larger than $\sim 10^8$ cm^{-3}. On the other hand, the presence of the strong intercombination lines of CIII] $\lambda 1909$, NIII] $\lambda 1750$, NIV] $\lambda 1486$ and OIII] $\lambda 1663$ suggest that the electron density is $\leq 10^{12}$ cm^{-3}.

From the recent observations, high local densities in the BLR ($n_H \sim 10^{11}$ to $10^{11.5}$ cm^{-3}) have been deduced from the study of UV Fe II emission of an extreme NLSy1 object, I Zw 1 [76], and in the quasars: LBQS 2113-4538 [77], CTS C30.10 [78], and HE 0435-4312 [79]. Using the intercombination line ratio Si III]/C III], [80] found the density of $10^{10.25}$ cm^{-3} from the study of UV absorbers of two NLSy1 galaxies. Using the photoionization modelling, [81] found the density of an intrinsic absorbing cloud in the bright quasar HS 1603 + 3820 to be of the order of $n_H \sim 10^{10}$ to 10^{12} cm^{-3}.

Constraining the density, and hence the location of the ionized gas remains one of the biggest challenge in the study of AGNs. Therefore it is important to look for the independent methods to constrain the gas density of the line emitting and absorbing region in AGNs. In this work, I utilize the observationally constrained values of different range of gas densities in the photoionization simulations of absorbing and emitting medium. I show that the variation in density significantly affects both: the AMD, and the line emission properties of ionized gas. Furthermore, I demonstrate that a comparison of the modelled AMDs and line emissivity profiles with that obtained from observations provide an independent method of constraining both: the gas density of WA, and line emitting regions in AGNs.

1.3 Objectives of the Thesis

The objective of this thesis is to use the photoionization simulations to study the physical conditions of the absorbing and emitting medium in AGNs. In this work, I used the publicly available photoionization code CLOUDY for the study of line emission regions, and the code TITAN to simulate the absorbing regions, in AGNs. The simulated results are used to derive the observable features such as the intermediate line emission and the absorption measure distribution. These physical properties are finally compared directly with that obtained from available observations published in the literature. In particular, the main objectives of this thesis are:

1. To show the importance of gas density in the photoionization modelling of an ionized gas in AGNs.
2. To investigate the physical conditions of the warm absorbing clouds in the Seyfert galaxy Mrk 509 and show that the constant total pressure cloud in photoionization computations reproduces the observed property, namely the absorption measure distribution.
3. To show that the absorption properties depend on the gas density for several AGNs.
4. To study the emissivity profiles of the major emission lines in AGN and to investigate the role of the dust in the extinction of the emission.
5. To show the existence of intermediate line region in several AGN types.
6. To study how intermediate line region depends on the density radial profile of considered clouds.
7. To consider realistic cloud density profile for radially distributed clouds, and to study how this profile affects the emission line luminosities.
8. Finally, to compare the results of my simulations with observational properties and estimate the density of the gas clouds, both in absorption and emission.

References

1. Lynden-Bell D (1969) Nature 223:690
2. Shakura NI, Sunyaev RA (1973) A&A 24:337
3. Hasinger G (2001) Quasars, AGNs and related research across 2000. In: Setti G, Swings J-P (eds) Conference on the Occasion of L. Woltjer's 70th Birthday, p 14
4. Fath EA (1909) Lick Obs Bull 5:71
5. Hubble EP (1926) ApJ, 64
6. Seyfert CK (1943) ApJ 97:28
7. Ho LC, Filippenko AV, Sargent WLW (1997) ApJS 112:315
8. Fanaroff BL, Riley JM (1974) MNRAS 167:31P
9. Antonucci R (1993) ARA&A 31:473
10. Urry CM, Padovani P (1995) PASP 107:803
11. Antonucci RRJ (1984) ApJ 278:499
12. Done C, Krolik JH (1996) ApJ 463:144
13. Collin S, Kawaguchi T, Peterson BM, Vestergaard M (2006) A&A 456:75

14. Radomski JT et al (2008) ApJ 681:141
15. Burtscher L et al (2013) A&A 558:A149
16. García-Burillo S et al (2016) ApJ 823:L12
17. Gallimore JF et al (2016) ApJ 829:L7
18. Malkan MA, Gorjian V, Tam R (1998) ApJS 117:25
19. Blandford RD, Znajek RL (1977) MNRAS 179:433
20. Rees MJ, Begelman MC, Blandford RD, Phinney ES (1982) Nature 295:17
21. Narayan R, Yi I (1995) ApJ 444:231
22. Zoghbi A, Fabian AC, Reynolds CS, Cackett EM (2012) MNRAS 422:129
23. Nandra K, Pounds KA (1994) MNRAS 268:405
24. Pounds KA, Nandra K, Stewart GC, George IM, Fabian AC (1990) Nature 344:132
25. Arav N, Brotherton MS, Becker RH, Gregg MD, White RL, Price T, Hack W (2001) ApJ
    546:140
26. Weymann RJ (1995) In: Meylan G (ed) QSO absorption lines, p 213
27. Crenshaw DM, Kraemer SB, Boggess A, Maran SP, Mushotzky RF, Wu C (1999) ApJ 516:750
28. Kaspi S et al (2001) ApJ 554:216
29. Behar E, Rasmussen AP, Blustin AJ, Sako M, Kahn SM, Kaastra JS, Branduardi-Raymont G,
    Steenbrugge KC (2003) ApJ 598:232
30. Steenbrugge KC, Kaastra JS, de Vries CP, Edelson R (2003) A&A 402:477
31. Turner AK, Fabian AC, Lee JC, Vaughan S (2004) MNRAS 353:319
32. Costantini E et al (2007) A&A 461:121
33. Chartas G, Brandt WN, Gallagher SC (2003) ApJ 595:85
34. Markowitz A, Reeves JN, Braito V (2006) ApJ 646:783
35. Dauser T et al (2012) MNRAS 422:1914
36. Tombesi F, Cappi M, Reeves JN, Nemmen RS, Braito V, Gaspari M, Reynolds CS (2013)
    MNRAS 430:1102
37. Laha S, Guainazzi M, Chakravorty S, Dewangan GC, Kembhavi AK (2016) MNRAS 457:3896
38. Kraemer SB, Tombesi F, Bottorff MC (2018) ApJ 852:35
39. Brotherton MS, Wills BJ, Francis PJ, Steidel CC (1994) ApJ 430:495
40. Mason KO, Puchnarewicz EM, Jones LR (1996) MNRAS 283:L26
41. Crenshaw DM, Kraemer SB (2007) ApJ 659:250
42. Hu C, Wang J-M, Ho LC, Chen Y-M, Bian W-H, Xue S-J (2008a) ApJ 683:L115
43. Hu C, Wang J-M, Ho LC, Chen Y-M, Zhang H-T, Bian W-H, Xue S-J (2008b) ApJ 687:78
44. Zhu L, Zhang SN, Tang S (2009) ApJ 700:1173
45. Li Z et al (2015) ApJ 812:99
46. Halpern JP (1984) ApJ 281:90
47. Collinge MJ et al (2001) ApJ 557:2
48. Kaastra JS, Steenbrugge KC, Raassen AJJ, van der Meer RLJ, Brinkman AC, Liedahl DA,
    Behar E, de Rosa A (2002) A&A 386:427
49. Netzer H et al (2003) ApJ 599:933
50. Krongold Y, Nicastro F, Brickhouse NS, Elvis M, Liedahl DA, Mathur S (2003) ApJ 597:832
51. Yaqoob T, McKernan B, Kraemer SB, Crenshaw DM, Gabel JR, George IM, Turner TJ (2003)
    ApJ 582:105
52. Blustin AJ et al (2003) A&A 403:481
53. Różańska A, Czerny B, Siemiginowska A, Dumont A-M, Kawaguchi T (2004) ApJ 600:96
54. Steenbrugge KC et al (2005) A&A 434:569
55. Winter LM, Mushotzky R (2010) ApJ 719:737
56. Winter LM, Veilleux S, McKernan B, Kallman TR (2012) ApJ 745:107
57. Laha S, Guainazzi M, Dewangan GC, Chakravorty S, Kembhavi AK (2014) MNRAS 441:2613
58. Holczer T, Behar E, Kaspi S (2007) ApJ 663:799
59. Behar E (2009) ApJ 703:1346
60. Detmers RG et al (2011) A&A 534:A38
61. Stern J, Behar E, Laor A, Baskin A, Holczer T (2014) MNRAS 445:3011

62. Adhikari TP, Różańska A, Sobolewska M, Czerny B (2016) In: Różańska A, Bejger M (eds) 37th meeting of the polish astronomical society, vol 3, pp 239–242
63. Kaspi S, Smith PS, Netzer H, Maoz D, Jannuzi BT, Giveon U (2000) ApJ 533:631
64. Peterson BM et al (2000) ApJ 542:161
65. Peterson BM et al (2013) ApJ 779:109
66. Netzer H, Laor A (1993) ApJ 404:L51
67. Crenshaw DM, Kraemer SB, Schmitt HR, Kaastra JS, Arav N, Gabel JR, Korista KT (2009) ApJ 698:281
68. Krolik JH, Kriss GA (1995) ApJ 447:512
69. Nicastro F, Fiore F, Perola GC, Elvis M (1999) ApJ 512:184
70. Kaastra JS et al (2012) A&A 539:A117
71. Arav N et al (2012) A&A 544:A33
72. Korista KT, Bautista MA, Arav N, Moe M, Costantini E, Benn C (2008) ApJ 688:108
73. Moe M, Arav N, Bautista MA, Korista KT (2009) ApJ 706:525
74. Bautista MA, Dunn JP, Arav N, Korista KT, Moe M, Benn C (2010) ApJ 713:25
75. Kaastra JS et al (2004) A&A 428:57
76. Bruhweiler F, Verner E (2008) ApJ 675:83
77. Hryniewicz K, Czerny B, Pych W, Udalski A, Krupa M, Świętoń A, Kaluzny J (2014) A&A 562:A34
78. Modzelewska J et al (2014) A&A 570:A53
79. Sredzinska J et al (2016) ArXiv e-prints
80. Leighly KM (2004) ApJ 611:125
81. Różańska A, Nikołajuk M, Czerny B, Dobrzycki A, Hryniewicz K, Bechtold J, Ebeling H (2014) New A 28:70

Chapter 2
Photoionization Simulations of AGN Environment

Abstract When the diffuse gas is photoionized by the optical/UV/X-ray photons, it gives rise to the emission/absorption lines from the irradiated region. Such effect is widely observed in different AGN components: BLR, ILR, NLR, WA and dusty torus, where we strongly believe that photoionized material is illuminated by the AGN nuclear continuum. The description of the photoionization process and the commonly adopted procedure in the calculations of the radiation transfer through the gas clouds is described in this chapter. Particularly, I describe the parameters and the theoretical formulations used in the photoionization simulations performed with cloudy and titan.

When the diffuse gas is photoionized by the optical/UV/X-ray photons, it gives rise to the emission/absorption lines from the irradiated region. Such effect is widely observed in different AGN components: BLR, ILR, NLR, WA and dusty torus, where we strongly believe that photoionized material is illuminated by the AGN nuclear continuum.

Since many years, a big theoretical effort was made to simulate the interaction of radiation with matter, but still existing codes work under certain assumptions. In this thesis, I made use of photoionization computations of ionization and thermal gas structure in order to put constraints on observational AGN features. I used the numerical codes CLOUDY versions C13.02 [1], later updated to C17 [2], and TITAN [3, 4]. Although both codes solve the transfer of radiation through the gas in thermal and ionization equilibrium with non-LTE equation of state, they do differ in radiative transfer treatment, assumptions about gas geometry, and the number of spectroscopic lines considered. In this Chapter, I describe the underlying assumptions and processes that are employed in the study of ionized gas, and display the most important differences between the numerical codes used in my research.

To make this Chapter clear, first I describe the most important equations in the photoionization approach. In the second step, I list boundary conditions, assumptions and definition of input parameters used in both codes. At the end of each Section, I

© Springer Nature Switzerland AG 2019   15
T. P. Adhikari, *Photoionization Modelling as a Density Diagnostic of Line Emitting/Absorbing Regions in Active Galactic Nuclei*, Springer Theses, https://doi.org/10.1007/978-3-030-22737-1_2

describe the differences in the numerical procedures adopted by TITAN and CLOUDY codes. In the last step, I emphasize what we can conclude by comparing results of photoionization modelling with current observations.

2.1 Radiative Transfer Problem—Set of Equations

To simulate the pass of radiation through the matter, the set of equations which bounds the radiation flow with the gas structure should be solved under several equilibrium conditions. Most photoionization codes put strong constraints on ionization and thermal equilibrium, but advanced computations often require also hydrostatic equilibrium to be achieved. In this thesis, I consider only one-dimensional computations of the radiative transfer.

In general, the following equations account for the full radiative transfer problem:

- Equation of radiative transfer for the specific intensity I_ν at frequency ν, in plane-parallel geometry on the monochromatic optical depth τ:

$$\mu \frac{dI_\nu}{d\tau} = -I_\nu + \frac{j_\nu}{\kappa_\nu + \sigma_\nu} = S_\nu - I_\nu, \qquad (2.1)$$

where S_ν is the frequency dependent source function, μ stands for the cosine of zenithal angle, and j_ν, κ_ν and σ_ν denote frequency dependent emission, absorption and scattering coefficients for 1 gram, respectively. The solution of this equation depends on the adopted gas structure. The above equation is intended to solve the transfer of radiation in a medium where the optical depth is counted in the same direction as that of the geometrical thickness dz, i.e. $d\tau = (\kappa_\nu + \sigma_\nu)\rho\,dz$, where ρ denotes gas density. In case of stellar atmospheres, τ is measured backward along the ray and it takes the form $d\tau = -(\kappa_\nu + \sigma_\nu)\rho\,dz$. More complicated gas structure requires more advanced numerical methods to solve this equation. Furthermore, the complexity of numerical solution depends on the considered atomic data which account for the opacity coefficients and directly increases computational time. The formal solution of the Eq. 2.1 can be written as

$$I_\nu(\tau_\nu, \mu) = I_\nu(0, \mu)e^{-\tau_\nu} + \int_0^{\tau_\nu} e^{-(\tau_\nu - \tau_\nu')/\mu} S_\nu(\tau_\nu')d\tau_\nu'/\mu. \qquad (2.2)$$

The first and second term of the Eq. 2.2 represent the initial intensity attenuated by the absorption and the integrated source function respectively.

- Ionization equilibrium: An ionization equilibrium stage is reached when there is a balance between the photoionization and the recombination rates occuring when photons interact with the atomic gas. As a result, the population of ions is reached to be stable and the ionic densities can be calculated. The ionization equilibrium equation for any two successive stage of ionization i and $i + 1$ of a given element X, can be written as

$$n(X^{+i}) \int_{\nu_i}^{\infty} \frac{4\pi J_\nu}{h\nu} a_v(X^{+i}) d\nu = n_e n(X^{+i+1}) \alpha_G(X^{+i}, T), \qquad (2.3)$$

where $n(X^{+i})$ and $n(X^{+i+1})$ are number densities of the two consecutive ionization states; $a_{v(X^{+i})}$ is the cross section at the ground state of X^i with threshold frequency ν_i, which is a minimum value of frequency for photon to ionize a given atom; n_e is the electron density and α_G is the recombination coefficient of the ground level of X^{+i+1} to all levels of X^{+i}. Equation 2.3 with the total number of ions at all stages of ionization

$$n(X) = n(X^0) + n(X^{+1}) + n(X^{+1}) + \cdots + n(X^{+n}) \qquad (2.4)$$

completely determines the ionization balance at each point. The mean intensity J_ν includes both the continuum and diffuse radiation contributions.

• Radiative (thermal) equilibrium: The thermal structure inside the slab of the gas cloud is obtained by solving for the electron temperature at which the thermal balance occurs. The local energy balance equation writes:

$$\int \frac{dF_\nu}{dz} d\nu = 0 = n_e n_H [\Lambda_{tot} - \Gamma_{tot}], \qquad (2.5)$$

where n_H is the hydrogen number density and F_ν is the monochromatic flux of the radiation. Λ_{tot} and Γ_{tot} are total cooling and total heating rates in erg cm^3 s^{-1}. All the important mechanisms: photoionization, recombination, Compton process, free-free process and line heating and cooling should be properly included in the total heating and cooling rates. Note that Compton heating-cooling in photoionization codes is treated globally and expressed by the standard formula

$$\Gamma_C - \Lambda_C = \frac{\sigma_T}{m_e c^2} \frac{1}{n_H} 4k(T_{IC} - T) \int 4\pi J_\nu d\nu, \qquad (2.6)$$

where Γ_C and Λ_C are Compton heating and cooling rates, and T_{IC} is the Compton temperature which depends on the spectral distribution of the local flux. m_e, c, T and σ_T are the mass of the electron, velocity of light, the gas temperature and the Thomson cross section respectively. Another important contributor to the cooling rate that can be described by analytical formula, is free-free radiation mechanism. The rate of cooling per unit volume Λ_{FF} due to free-free process is approximately written as

$$\Lambda_{FF} = 1.42 \times 10^{-27} T^{1/2} Z^2 n_e n_+ g_{ff}, \qquad (2.7)$$

where Z is the charge of the ions and n_+ is the number density of the ions. g_{ff} is the mean Gaunt factor of the free-free emission. All other bound-free and bound-bound processes have to be treated numerically by adding contributions from each ion population, therefore it is not possible to present one formula. Therefore, the contribution of all cooling and heating processes for an exemplary cloud computed by TITAN code is presented in Fig. 2.1.

The overall shape of different processes across the gas, which determines local gas temperature, depends directly on the amount of cross sections on different transitions included in the numerical algorithm. Photoionization codes differ by the amount of atomic data taken into account. Some of those data appropriate for heavy elements are still not determined in laboratories.

- Hydrostatic equilibrium equation generally assumes that only gas and radiation pressure (P_{gas}+ P_{rad}) account for the total pressure which has to balance the gravity force g. In the theory of stellar atmospheres, it is generally written as:

$$\frac{dP_{gas}}{d\tau} = \frac{g}{\kappa_\nu + \sigma_\nu} - \frac{dP_{rad}}{d\tau}. \tag{2.8}$$

Photoionization codes used for this research do not take into account the gravity force. The most common and computationally easiest approach is to assume that the illuminated cloud is under constant density, hereafter CD models. In such cases, hydrostatic equilibrium is not needed at all, but the gas cloud is not influenced by the gravitational force of SMBH. Numerically more complicated but physically more consistent is the approach that illuminated cloud is under constant total pressure, hereafter CP models [5]. Even if the gravity force is still neglected, the gas cloud is subjected to the AGN radiation field, and above differential equation has to be solved. The limitation of the solution of hydrostatic equilibrium in case of photoionized clouds in AGNs is presented in the Sect. 3.3.

2.1.1 TITAN

TITAN numerical code was mainly developed to compute the temperature, ionization structure and the emission spectrum of a thick hot slab of gas photoionized by the radiation from the central source. Assuming the local balance between the ionization and recombination of ions, excitation and de-excitation, and most important atomic line transitions, TITAN computes a physical state of the gas at each depths in thermal and hydrostatic equilibrium. About 4000 line transitions are included in TITAN code.

TITAN uses Accelerated Lambda Iteration (ALI) method to solve the radiative transfer, which takes into account the proper source function at each depth of the cloud and computes line and continuum intensity self consistently. I refer the reader to the paper by Dumont et al. [6] where a detailed description of the ALI method is done in the context of the X-ray absorber. At the cost of computing time, the structure of the gas is obtained iteratively until the convergence is reached.

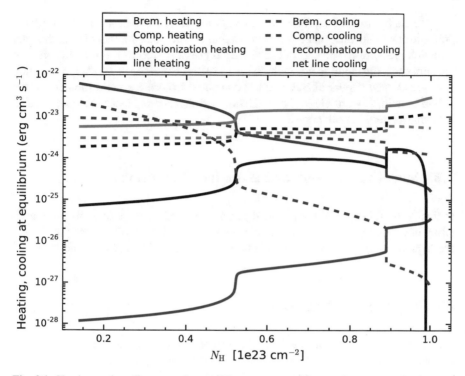

Fig. 2.1 Heating and cooling rates due to different processes for a reference gas cloud model photoionized by the typical AGN radiation field

2.1.2 CLOUDY

CLOUDY numerical code is public, regularly updated, and used by the astronomers to study the photoionized plasma in many different environments such as: clouds in the interstellar medium (ISM), AGNs and planetary nebulae. This code is very user friendly due to its richness in the documentation containing extensive explanation of the physics of the plasma subjected to different perturbations (the most important are the Hazy1 and Hazy2[1] documentations).

CLOUDY uses escape probability method of the radiative transfer computation across depth of the photoionized clouds. This method is reliable and useful for the optically thin clouds. However, for the optically thick clouds this is less accurate than the ALI method as it does not take into account the proper source function which depends on the diffuse radiation inside the cloud. But thanks to this simplification, the CLOUDY code collects the biggest amount of atomic data, including dust opacities, among all photoionization codes available in astrophysics.

---

[1] http://nublado.org/.

Also the hydrostatic equilibrium is simplified in CLOUDY in a way that only the pressure due to the attenuated incident continuum and lines are included in the total radiation pressure term. The initial radiation pressure due to the radiation flux at the surface of the cloud is not included in CLOUDY. However in TITAN the radiation pressure at each depths of the cloud is computed from the second moment of the radiation field. Nevertheless, the ionization and thermal balances are computed in the same manner in both codes.

2.2 Model Parameters and Boundary Conditions

In this Section, I describe the physical parameters used in the photoionization modelling of the emitting and absorbing plasma studied in this thesis. These parameters are required as an input parameters in both numerical codes, CLOUDY and TITAN.

2.2.1 Ionization Parameter

When the gas cloud is illuminated by the incident radiation field, it is ionized across the depths. The strengths of ionization is parametrized by the parameter ξ [erg cm s^{-1}] commonly known as the ionization parameter. In the literatures, various forms of the ionization parameters are defined. In this thesis, I use the ionization parameter defined at the surface of the cloud by the expression (unless otherwise stated)

$$\xi = \frac{L_{ion}}{n_H r^2},\qquad(2.9)$$

where L_{ion} is the hydrogen ionizing luminosity i.e., the luminosity integrated between $1-1000$ Ryd and r is the distance to the cloud from the illuminating source.

In the TITAN code, the ionization parameter is self consistently computed at each depths of the gas cloud using the expression

$$\xi = 4\pi c k T \frac{P_{rad}}{P_{gas}} = 4\pi c \frac{P_{rad}}{n_H},\qquad(2.10)$$

where k is the Boltzmann's constant. P_{rad} and P_{gas} are computed self consistently by TITAN at each layer of the photoionized gas cloud.

Often for visualization, the gas dynamic behaviour of the thermal and ionization properties is traced by defining the dynamical ionization parameter Ξ which relates to the parameter ξ by the following equation

$$\Xi = \frac{\xi}{4\pi c k T} = \frac{L_{ion}}{4\pi c r^2}\frac{1}{n_H k T} = \frac{F_{ion}}{c P_{gas}} = \frac{P_{rad}(r)}{P_{gas}(r)},\qquad(2.11)$$

where F_{ion} is the ionizing flux of the incident radiation. Here, I note that we do not follow the standard convention that P_{gas} accounts only for hydrogen density number n_H. While computing the gas pressure for a fully ionized gas with heavy element abundance, the total density number $\approx 2.3\, n_H$ should be taken into account, as it is in the basic definition of Ξ, given by Krolik et al. [7]. In this thesis, I self consistently compute the ionization parameter and its variation with the depth of the cloud from the illuminated face using Eq. 2.11. This is possible only with TITAN code where radiation pressure is calculated at each cloud depth, from the second moment of the radiation field.

While dealing with the emission regions in Chap. 5, I use the dimensionless ionization parameter U defined in Eq. 2.12, by the ratio of hydrogen ionizing photon density to the total hydrogen density.

$$U = \frac{Q_H}{4\pi r^2 n_H\, c},$$

(2.12)

where Q_H is the number of hydrogen ionizing photons per second and it is obtained by integrating the photon fluxes in the energy range 1–1000 Ryd.

2.2.2 SED Shape

One of the most important parameter in any consistent photoionization computations of the irradiated medium is the broad band spectral energy distribution (SED) of the source. It has been shown that the shape of the SED affects the ionization and thermal structure of the photoionized gas [8]. The authors found that the stability curve obtained with the SED dominated by hard X-ray power law differs significantly from that obtained with the SED dominated by the soft component (i.e., accretion disk component). This is connected with the fact that the dominant heating/cooling mechanisms are different for the different SED cases: bremsstrahlung is the dominant cooling in the cloud irradiated by the SED with significant soft component while Compton scattering is responsible for high temperature equilibrium in case of SED with hard X-ray power law. Moreover, Chakravorty et al. [9] have shown that the nature of the stability curve strongly depends on the shape of the spectral radiation used in the photoionization computations. In particular, the authors discussed the effect of soft X-ray excess on the overall shape of the stability curve.

I use various SED shapes described in Chaps. 4 and 5, to investigate how the photoionization simulations of the absorbing and emitting medium in AGN depend on the shape of the radiation field.

2.2.3   Gas Density, Column Density and Pressure

Gas density is another important parameter in the simulations of the photoionized gas. Throughout this thesis, I denote the gas density at the surface of the gas clouds by $n_{\rm H}$. In the CD case, it remains the same across the depth of the cloud. However in the CP assumptions, the density is stratified across the inner layers in the increasing order. This is due to the fact that, as the radiation passes through the slab of the material it gets absorbed and partially reprocessed and thereby increases the gas pressure. This causes the rise in the gas density. Nevertheless, the sum of the radiation pressure and the gas pressure is the same in each layer. I note here, that in this work, the magnetic pressure is not taken into account. However, the pressure due to the turbulence is taken into account whenever I use the turbulent velocity in our computations. The stratification of the pressure and density across the depths of a single cloud and its comparison with the CD cases are discussed in details in the Sect. 3.3.

The total thickness of the cloud is set by assigning the column density, denoted by $N_{\rm H}$, in such a way that $dN_{\rm H} = n_{\rm H}\, dz$, where dz is the geometrical thickness of an individual layer. Column density is always an input parameter in photoionization calculations.

2.2.4   Geometry

In all the photoionization computations performed in this thesis, I use the open geometry of the gas clouds. This means, all radiation that escape from the illuminated face of the cloud back towards the source of continuum radiation, do not further interact with gas. This differs from the closed geometry in which the radiation that goes in the backward direction interacts with gas (for example: the case of planetary nebula illuminated by star located in its center).

In the open geometry, the CLOUDY code allows for three options depending on the thickness of the gaseous cloud. So, if the ratio $\frac{\Delta r}{r} < 0.1$ (where Δr is the thickness of the shell), the geometry is plane parallel, whereas, if $0.1 < \frac{\Delta r}{r} < 0.3$, then it will be thick shell geometry. And it becomes spherical when the shells of gas clouds are extended, i.e., $\frac{\Delta r}{r} \geq 3$, consequently the radiation pressure falls faster as we go deeper into the cloud i.e., $P_{\rm rad} \propto 1/r^2$ because of the geometrical dilution and the absorption by ionized gas. This causes the decrease in ionization parameter and hence the temperature of the gas decreases even for the constant density at each cloud zones.

Throughout this thesis, a plane parallel geometry is adopted in both codes TITAN and CLOUDY unless otherwise stated.

2.2.5 Chemical Composition

In any photoionization computations, chemical abundances are very essential for the correct modelling of the interactions of photons with individual ions. Measurement of the chemical composition of the AGN clouds is difficult because of the nonstellar nature of the continuum shape. However, some line ratios that are independent of the temperature and ionization parameter of the cloud zones can be used to estimate the composition. The ratio of semi-forbidden line ratio NIV λ1486 to CIV λ1549 can be used as a good C/N abundance indicator. Nevertheless, there exist big uncertainties in the derived abundances and the process itself.

Only few sources in the literatures are available where a detailed study of the abundances in the AGNs are done. One of the most studied source in which the metal abundances are derived from the observations is Mrk 509. Costantini et al. [10] showed that the best fitted abundances for Mrk 509 are Solar. This is one of the few cases of metallicity estimations for the low redshift AGNs. For the high redshift quasars, the abundances estimations are usually done using intercombination line ratios: N III]/O III], N V/He II, N IV]/O III] and N IV]/C IV. From this method the typical metallicities in the quasar environment are $1-5$ times the Solar metallicity (Hamann et al. [11], and references therein). For NGC 5548, Steenbrugge et al. [12] considered relative abundances of a few elements to produce alternative photoionization model to test the model of X-ray absorption. A similar approach was used in Crenshaw et al. [13], but none of those authors have derived abundances from the data.

As described in the previous Sections, photoionization calculations already require large number of input parameters: the shape of illuminated radiation, the ionization parameter and several global gas physical parameters. Therefore, it is a wise decision to choose one particular assumption about the metallicity in any photoionization computations since the final result is already dependent on many other parameters. In this thesis, I mostly use the *Solar abundances* as shown in the column 3 of the Table 2.1 in the computed models of absorption and emission. At the moment, TITAN uses only 10 of the most abundant elements: H, He, C, N, O, Ne, Mg, Si, S and Fe. This differ from the case of CLOUDY where all 30 elements shown in the Table 2.1 are included by default. However, when I compare the results between TITAN and CLOUDY computations in Chap. 3, the rest of the elements in the CLOUDY are turned off and the same values are set for the common 10 most abundant elements in both codes.

At the nearby distances from the central engine of the AGN, the temperature is high enough to sublimate the dust grains and hence it is no longer present there. However, at larger radii, the temperature is low enough for the dust grains to survive. So the presence of the dust in the gas clouds above certain distance called sublimation radius should be properly accounted. Nevertheless, including the dust grains in the photoionization computations is not so simple as grains are made up of elements that are condensed from the gas phase. It would be inconsistent to assume *Solar abundances* for all elements and also include grains. This is because in reality, certain

Table 2.1 The chemical abundances of the first 30 elements used in the photoionization computations. The third colum lists the *Solar abundances* described in the Hazy1 documentation of the CLOUDY and taken from the paper by Grevesse and Anders [14]. The fourth column shows the *ISM abundances* used in the CLOUDY

Atomic number	Element	Solar abundances	ISM abundances
1	Hydrogen H	1.00E00	1.00E00
2	Helium He	1.00E–01	9.80E–2
3	Lithium Li	2.04E–09	5.40E–11
4	Beryllium Be	2.63E–11	2.63E–11
5	Boron B	7.59E–10	8.90E–11
6	Carbon C	3.55E–04	2.51E–04
7	Nitrogen N	9.33E–05	7.94E–05
8	Oxygen O	7.41E–04	3.19E–04
9	Fluorine F	3.02E–08	2.00E–08
10	Neon Ne	1.17E–04	1.23E–04
11	Sodium Na	2.06E–06	3.16E–07
12	Magnesium Mg	3.80E–05	1.26E–05
13	Almunium Al	2.95E–06	7.94E–08
14	Silicon Si	3.55E–05	3.16E–06
15	Phosphorus P	1.35E–07	1.60E 07
16	Sulphur S	1.62E–05	3.24E–05
17	Chlorine Cl	1.88E–07	1.00E–07
18	Argon Ar	3.98E–06	2.82E–06
19	Potassium K	1.35E–07	1.10E–08
20	Calcium Ca	2.29E–06	4.10E–10
21	Scandium Sc	1.58E–09	1.00E–20
22	Titanium Ti	1.10E–07	5.80E–10
23	Vanadium V	1.05E–08	1.00E–10
24	Chromium Cr	4.84E–07	1.00E–08
25	Manganese Mn	3.42E–07	2.30E–08
26	Iron Fe	3.24E–05	6.31E–07
27	Cobalt Co	8.32E–08	5.90E–09
28	Nickel Ni	1.76E–06	1.82E–08
29	Copper Cu	1.87E–08	1.50E–09
30	Zinc Zn	4.52E–08	2.00E–08

elements, especially Ca, Al, Ti, and Fe, are strongly depleted from the gas phase in the ISM and are believed to be present as grains. In the study of emission line properties in the Chap. 5 I use both the Solar composition and ISM with grains whenever it is necessary. Abundances used in this thesis with CLOUDY code are given in column 3 and 4 of Table 2.1. In the study of absorbing clouds in the Chap. 4, I do not include the contribution from the dust grains in the total chemical composition due to high gas temperature.

2.3 What Outputs of Photoionization Simulations Can We Use to Compare with Observations?

The thermal and ionization structures computed in the photoionization simulations are further utilized to estimate the physical quantities and their distributions that can be directly compared with observations. In this Section I present such outputs obtained from the simulations done in this thesis.

2.3.1 Stability Curve Analysis

The stability of the irradiated gas clouds is generally studied using the diagram obtained by plotting the temperature of the gas as a function of the dynamical ionization parameter Ξ defined as the ratio of radiation pressure to the gas pressure (see Eq. 2.11). Each point of the curve gives the temperature and corresponding value of the ionization parameter, both obtained by solving the physical equations assuming thermal balance i.e., total heating is equal to the total cooling along the curve. The part of the curve where the slope is positive gives the stable part of the absorber whereas the part with negative slope represents the thermally unstable region.

In literatures, there have been different forms of stability curves in practice [8, 15–17], though they differ only by some constants but the resulting nature is very similar. The stability curve is usually computed by simulating a grid of constant density clouds located at the same distance from the illuminating radiation source. When the cloud density changes, it also changes the ionization parameter and hence the resulting gas temperature also varies. However, the constant pressure assumption in the photoionized single gas cloud self consistently allows to compute the stability curve and study how the temperature changes with the ionization degree of the gas. In this work, I follow the standard convention of defining the stability curve by plotting the temperature against the dynamical ionization parameter in both cases of CD and CP assumptions. I discuss the comparison of the stability curves obtained from CLOUDY and TITAN codes for CD and CP assumptions in the Sect. 3.4.

2.3.2 Application to AMD

The photoionization simulations are very useful to study X-ray absorption in AGNs allowing to constrain the gas properties by comparing the results of the simulations to that of the observations. High resolution X-ray data from observations are used to constrain the properties like the absorbing column and the ionization degrees of the gas. The measurement of the column density of the absorber as a function of ionization degree at different radial location of the outflow is defined as the absorption measure distribution (AMD).

From observations it has been shown that the ionization degree varies by ~ 4 orders of magnitude for many AGNs [18–20]. The measurement of the AMD provides the information about the ionization and temperature stratification of outflow in AGNs. The expression for AMD was first formulated by Holczer et al. [18] as

$$\text{AMD} = \frac{dN_{\text{H}}}{d(\log \xi)} = 2.303\, \xi \frac{dN_{\text{H}}}{d\xi}, \tag{2.13}$$

where, ξ is the ionization parameter defined in Eq. 2.9. Total hydrogen column density N_{H}, in the above equation is defined along the line of sight. Since we do not see hydrogen absorption lines in the spectra it is not possible to measure the hydrogen ionic column density directly. The N_{H} is derived by using the ionic column densities of each observed heavy element ions. The ionic column densities of all the relevant ions are obtained by fitting the corresponding absorption lines in the observed spectra of AGNs. The ionic column density is related to the AMD by an expression

$$N_{\text{ion}} = A_{Z_\odot} \int \text{AMD}\ f_{\text{ion}}(\log \xi)\ d(\log \xi), \tag{2.14}$$

where f_{ion} is the fractional ion abundance with respect to the total abundance of its element, $A_{Z_\odot}$ is the given element abundance relative to its Solar value. Assuming that fractional ion abundance peaks at the maximum value of ionization parameter, ξ_{max}, combination of Eqs. 2.13 and 2.14 gives

$$N_{\text{H}} \approx \frac{N_{\text{ion}}}{f_{\text{ion}}(\xi_{\text{max}}) A_{Z_\odot}}, \tag{2.15}$$

which was first used to derive AMD from high resolution X-ray data by Holczer et al. [18]. When ionic column densities are obtained from individual line fitting, Eq. 2.15 combines them with N_{H} through the photoionization calculations which provide us with the correct value of ξ_{max}. However, in reality the gas is distributed over the range of ξ and one must take into account the full dependence of f_{ion} on ξ. This makes the whole fitting procedure even more complicated. Therefore, any photoionization calculations made for this purpose have assumed thin CD slabs.

Note that, the assumption of thin photoionization slab is equivalent to the assumption that we are on the linear part of curve of growth and the observed ionic column

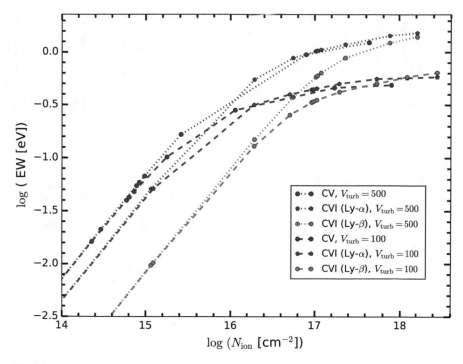

Fig. 2.2 Modelled curves of growth for from helium (red) and hydrogen-like (blue, green) Carbon ions. Saturation is visible in all cases starting at $N_{ion} = 10^{15}$ cm^{-2} for red and $N_{ion} \sim$ few 10^{16} cm^{-2} for blue lines. It is much stronger for lower velocity of 100 km/s (dashed lines), than for 500 km/s (dotted lines)

densities are proportional to the line equivalent widths (EW). Nevertheless, it may not be the case. Figs. 2.2 and 2.3, I present the results of modelled ionic column densities from TITAN calculations, where the curve of growth of helium and hydrogen-like ions of carbon and oxygen are shown for typical photoionization run. Two cases of turbulent velocities 100–500 km s^{-1} are shown.

It is obvious from these Figs. 2.2 and 2.3 that typical values of measured ionic column densities (in some cases up to 10^{18} cm^{-2}) are lying on the exponential part of the curve of growth, which means lines should be saturated, and the assumption of "thin slab" is not valid any longer. Furthermore, the effect of turbulent velocity on the line saturation is also clearly visible i.e., the saturation is much strong for the lower turbulent velocity in comparison to the higher velocity. Therefore, the derived ionic column densities from observations using thin slabs give only lower limits, and in reality can be one order of magnitude higher. Nevertheless, according to our knowledge, saturated lines were never reported from X-ray data. Furthermore, photoionization calculations of thick slabs make the whole AMD fitting procedure extremely complicated.

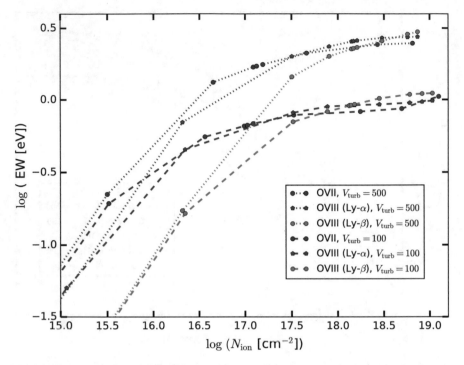

Fig. 2.3 Similar to Fig. 2.2 but for the ions of Oxygen atom

Since, high resolution X-ray data are available only for several objects, only few AMDs are already published. All of them are presented in Fig. 2.4 where the data were kindly provided for us by Ehud Behar.

Using the method described above, Holczer et al. [18] studied the AMD of the outflow in bright type 1 quasar IRAS 13349+2438 using *Chandra* X-ray spectrum obtained with high energy transmission grating spectrometer (HETGS) and found the distribution spans over four orders of magnitude with the absence of absorbing column around the ionization parameter $0.75 < \log (\xi \, [\mathrm{erg\,cm\,s^{-1}}]) < 1.75$. Behar [19] derived AMDs for five nearby Seyfert galaxies: NGC 3783, IRAS 13349+2438, MCG-6-15-30, NGC 5548 and NGC 7469 using the high resolution X-ray spectra from HETGS on board in *Chandra* and RGS spectrometer on board in *XMM-Newton*. In all AMDs, they found a gap at intermediate ξ values, i.e., in between $\log \xi \sim 0.8$ and 1.7 erg cm s^{-1}. The gap in the AMD distribution is interpreted as a thermally unstable region in the AGN outflow. However, for the Sy1.5 Mrk 509, using the 600 ks RGS data, Detmers et al. [20] obtained a AMD with two prominent minima located at $\log \xi \sim$ around 2–3 and 3–4, a different result that was obtained for five sources by Behar [19]. Laha et al. [21] concluded that AMD has a universal shape across all observed sources.

AMD modelling is very useful in constraining the physical properties of the absorbing gas in the AGN outflow. By comparing the modelled distribution of

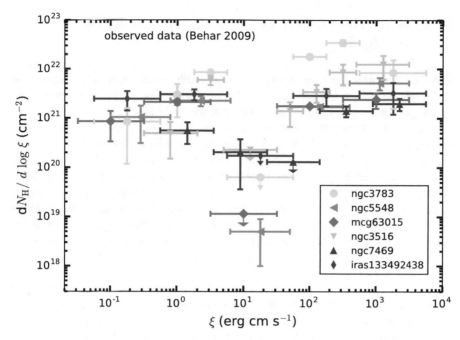

Fig. 2.4 The observed AMDs for six Seyfert galaxies plotted from Behar [19] with the authors permission. ©AAS. Reproduced with permission

absorbing column with that derived from the observed X-ray spectra using the high resolution spectrometer on board in space, one can infer the properties of the ionized gas. The first AMD models were computed by Stern et al. [22] with the numerical code CLOUDY assuming the gas clouds to be radiation pressure confined (RPC). Although the authors were successful in reproducing the observed AMD normalization, their RPC models do not exhibit the observed gap in the distribution. In RPC computations, the total pressure is dominated by the radiation pressure of the incident radiation (i.e., $P_{rad} \gg P_{gas}$) and is transferred to the gas as it passes through the plasma. So, at each depth of the gas cloud, the decrease in radiation pressure is contributing to the gas pressure. All other contribution to the total pressure term i.e., magnetic pressure, turbulent pressure and the pressure induced by the trapped line photons were ignored in the RPC computations. In the Sect. 3.3, I show that the RPC is one special solution of CP cloud models.

I used TITAN numerical code for the computation of the thermal and ionization structure of the gas ionized by an incident radiation of the central source of the Sy 1 Mrk 509 to obtain the AMD model for this source. Additionally, I investigated how the AMD shape changes with the shape of the incident SEDs. The purpose of doing this was mainly to explain the observed variation in the number of AMD minima and AMD normalization. I devote the Chap. 4 to present my research on AMD modelling.

2.3.3   Application to Emission Line Regions: BLR, ILR and NLR

The observed emission lines in AGN are most likely produced from the interaction of an intense non-stellar continuum radiation with the surrounding materials i.e., by the process of photoionization. The direct evidence to support this idea of line production is a correlation between the line and the continuum variations in many AGNs, where the continuum luminosity changes are followed by the line luminosity changes. The optical/UV emission lines present in the observed AGN spectra are key to understand the physical properties of the line emitting medium. The general picture derived from the observed spectra consists of two distinct regions; the dense BLR close to the AGN center and the rarer NLR located much further away. These two regions are well separated by the transition region without significant emission. However in the recent studies of several AGNs, the intermediate component of line emission originated in the transition region between BLR and NLR is also observed (see Sect. 1.1.3 for details).

Photoionization calculations provide a powerful tool to model the line emission from gas clouds powered by the continuum radiation from the nucleus. It has been frequently applied in the study of BLR and NLR in AGNs (e.g. [23–26]). The study of global properties of emission line regions in AGNs was done by Netzer and Laor [24] where the authors calculated emission from a continuous radial distribution of clouds extending from the BLR to the NLR, using the numerical photoionization code ION [27] (similar to the CLOUDY code). The clouds were assumed to be distributed above an AD, and their basic parameters have been changed with radial distance as

$$n_H \propto r^{-3/2}, \quad N_H \propto r^{-1}. \tag{2.16}$$

Their assumption was that the dust is present in NLR but sublimes in a higher ionization region, around the BLR, at the sublimation radius. The authors have shown that the reduced line emission vs radius is a result of the dust extinction of ionizing radiation as well as the dust destruction of the line photons.

In the frame of their model, the dust absorption becomes most efficient exactly between BLR and NLR giving rise to an empty intermediate region, where gas is present but the line emission is heavily suppressed by dust. The dust fully sublimates at smaller radii, and line emission increases dramatically, by about an order of magnitude, giving rise to the BLR. Netzer and Laor [24] successfully demonstrated the apparent gap in the line emission vs radius, although their result was obtained for a particular spectral energy distribution (SED) typical for Sy1 AGN [28], and for a specific value of the density $n_H = 10^{9.4}$ cm^{-3} at the sublimation radius at $\sim$0.1 pc. Furthermore, they have considered only one slope of power law density radial profile, and only one value of canonical AGN luminosity 10^{45} erg/s.

Dust is the basis of the AGN unification theory which proposes that all AGNs surrounded by optically thick dusty torus are fundamentally the same object observed from different inclination angles [29, 30]. It has been argued that the BLR clouds are free of dust since there is no depletion of refractory elements in the gaseous phase [31]. Usually, it is believed that, the radiation energy is so high in BLR, that the dust, if present, sublimes and no longer survives there [32]. On the other hand, observations have shown that the dust exists in NLR for almost all types of galaxies.

The presence of dust complicates the radiative processes by introducing additional physical mechanisms that influences the radiation matter interactions which has to be treated appropriately in the simulations of the gas emission ([33–35], and references therein). RM studies show that the BLR clouds are located at a distance smaller by a factor 4 to 5 than the hot dust emission [36, 37].

On the other hand, Nenkova et al. [38] has shown that the closest region where dust can survive in the full radiation field is the face of the dusty torus, located approximately at 0.4 pc for AGN of typical luminosity, given by the expression

$$R_d = 0.4\sqrt{L/10^{45}} \quad [\text{pc}]. \tag{2.17}$$

The position of sublimation radius strongly depends on the detailed dust composition which is still under discussion [39, 40]. The dust is most likely a mixture of amorphous carbon [41], silicate [42] and graphite grains [43]. The sublimation radius derived by Nenkova et al. [38] (see Eq. 2.17) corresponds to the temperature of sublimation of silicate grains only [44]. Graphite grains sublimate at larger temperature up to $\sim$2000 K [43, 44] and R_d takes the form

$$R_d = 0.06\sqrt{L/10^{45}}. \tag{2.18}$$

Nevertheless, AGN extinction curves do not show the 2175Å carbon feature [45] which makes the dust in the circumnuclear region of AGNs being different from Galactic ISM.

With the recent observations, the obvious question that should be answered is: What is the physical mechanism leading to the existence of ILR in some sources but not in all of them? In Chap. 5, I present the photoionization simulations of radially distributed clouds, following the approach by Netzer and Laor [24]. Nevertheless, in contrast to these authors, I show the existence of intermediate line region is AGNs of different types, which is in agreement with recent observations. I show that high density gas in the emitting regions successfully produces ILR region in all considered sources. Furthermore, I search how this result depends on other input parameters employed in photoionization simulations.

2.4 Conclusions

In this Chapter, I presented the underlying assumptions and the parameters that determine the physics of the photoionized gas clouds. My conclusions from this Chapter are listed as follows:

1. Although the numerical codes: TITAN and CLOUDY are comparable in many aspects in simulating the photoionized gas clouds, both codes differ in the treatment of radiative transfer, hydrostatic balance and the atomic databases used. Ionization balance and non-LTE equation of state are treated in the same way.
2. Heating and cooling processes of photoionized gas clouds can be analyzed after each numerical run. In this way, we can indicate the driving mechanism which is responsible for the cloud heating. In the following Chapter we show that there are at least two processes which keep cloud in high temperature equilibrium.
3. The assumption of a thin slab of gas used in the estimation of ionic column densities from X-ray observations may be invalid when observed lines are saturated. In case of strong line saturation such assumption provides only the lower limits for ionic column density derivation. Using the thick slab with the proper account of line saturation can result the ionic column densities higher, but the final value depends on gas turbulent velocity.
4. The presence of different dust grain components in AGNs is not yet a resolved issue. Depending on the mixture of the dust grain components the sublimation radius may differ. The presence/absence of dust should be properly taken into account in the simulations of the global emission line properties in AGNs (see Chap. 5 for details).

References

1. Ferland GJ et al (2013) Rev Mex Astron Astrofis 49:137
2. Ferland GJ et al (2017) Rev Mex Astron Astrofis 53:385
3. Dumont A-M, Abrassart A, Collin S (2000) A&A 357:823
4. Collin S, Dumont A-M, Godet O (2004) A&A 419:877
5. Różańska A, Goosmann R, Dumont A-M, Czerny B (2006) A&A 452:1
6. Dumont A-M, Collin S, Paletou F, Coupé S, Godet O, Pelat D (2003) A&A 407:13
7. Krolik JH, McKee CF, Tarter CB (1981) ApJ 249:422
8. Różańska A, Kowalska I, Gonçalves AC (2008) A&A 487:895
9. Chakravorty S, Kembhavi AK, Elvis M, Ferland G (2009) MNRAS 393:83
10. Costantini E et al (2016) A&A 595:A106
11. Hamann F, Warner C, Dietrich M, Ferland G (2007) Astronomical society of the pacific conference series, vol 373, The central engine of active galactic nuclei, Ho LC, Wang J-W, eds, p 653
12. Steenbrugge KC et al (2005) A&A 434:569
13. Crenshaw DM, Kraemer SB, Schmitt HR, Kaastra JS, Arav N, Gabel JR, Korista KT (2009) ApJ 698:281
14. Grevesse N, Anders E (1989) American institute of physics conference series, vol 183, Cosmic abundances of matter, Waddington CJ, ed, pp 1–8
15. Krolik JH, Kriss GA (1995) ApJ 447:512
16. Chakravorty S, Misra R, Elvis M, Kembhavi AK, Ferland G (2012) MNRAS 422:637
17. Dyda S, Dannen R, Waters T, Proga D (2017) MNRAS 467:4161
18. Holczer T, Behar E, Kaspi S (2007) ApJ 663:799
19. Behar E (2009) ApJ 703:1346
20. Detmers RG et al (2011) A&A 534:A38
21. Laha S, Guainazzi M, Dewangan GC, Chakravorty S, Kembhavi AK (2014) MNRAS 441:2613

22. Stern J, Behar E, Laor A, Baskin A, Holczer T (2014) MNRAS 445:3011
23. Netzer H (1990) Active galactic nuclei, Blandford RD, Netzer H, Woltjer L, Courvoisier TJ-L, Mayor M, (eds), pp 57–160
24. Netzer H, Laor A (1993) ApJ 404:L51
25. Dopita MA, Groves BA, Sutherland RS, Binette L, Cecil G (2002) ApJ 572:753
26. Baskin A, Laor A, Stern J (2014) MNRAS 438:604
27. Rees MJ, Netzer H, Ferland GJ (1989) ApJ 347:640
28. Mathews WG, Ferland GJ (1987) ApJ 323:456
29. Antonucci R (1993) ARA&A 31:473
30. Urry CM, Padovani P (1995) PASP 107:803
31. Gaskell CM, Shields GA, Wampler EJ (1981) ApJ 249:443
32. Czerny B, Hryniewicz K (2011) A&A 525:L8
33. Ferland G, Netzer H (1979) ApJ 229:274
34. Baldwin JA, Ferland GJ, Martin PG, Corbin MR, Cota SA, Peterson BM, Slettebak A (1991) ApJ 374:580
35. van Hoof PAM, Weingartner JC, Martin PG, Volk K, Ferland GJ (2004) MNRAS 350:1330
36. Suganuma M et al (2006) ApJ 639:46
37. Koshida S et al (2014) ApJ 788:159
38. Nenkova M, Sirocky MM, Nikutta R, Ivezić Ž, Elitzur M (2008) ApJ 685:160
39. Gaskell CM (2017) MNRAS 467:226
40. Xie Y, Li A, Hao L (2017) ApJS 228:6
41. Czerny B, Li J, Loska Z, Szczerba R (2004) MNRAS 348:L54
42. Lyu J, Hao L, Li A (2014) ApJ 792:L9
43. Baskin A, Laor A (2018) MNRAS 474:1970
44. Laor A, Draine BT (1993) ApJ 402:441
45. Maiolino R, Marconi A, Salvati M, Risaliti G, Severgnini P, Oliva E, La Franca F, Vanzi L (2001) A&A 365:28

Chapter 3
The Role of Gas Density

Abstract The preliminary photoionization simulations of gas illuminated by radiation field have shown that for a given L_{ion}/ξ ratio, it is hard to find overall differences
in transmitted spectrum for large range of gas densities. Only the line ratios of particular transitions may put additional constraints on the value of gas density and
therefore on the wind location. This result has led researchers to assume constant
density (CD) clouds in any photoionization calculations. However, in the recent
modelling of absorbing as well as emission gas in AGN environment, the clouds in
hydro-static equilibrium are also used. In this Chapter, I investigated the influence
of gas density on the overall structure of CD and constant pressure CD cloud. A
detailed comparison between CD and CP individual cloud is done, and all possible
solutions of hydro static equilibrium are considered.

The preliminary photoionization simulations of gas illuminated by radiation field
have shown that for a given L_{ion}/ξ ratio, it is hard to find overall differences in
transmitted spectrum for large range of gas densities. Only the line ratios of particular
transitions may put additional constraints on the value of gas density and therefore
on the wind location. This result has led researchers to assume constant density
(CD) clouds in any photoionization calculations. Keeping gas density constant over
the whole cloud allows us to skip solving the hydrostatic equilibrium, which makes
the numerical procedure simpler and more effective, also in CPU time. Therefore,
the most important input parameter in photoionization calculations is the ionization
parameter (Eq. 2.9 or equivalently Eq. 2.12).

In the recent modelling of absorbing as well as emission gas in AGN environment,
the clouds in hydrostatic equilibrium are also used [1–4]. The assumption of hydrostatic equilibrium requires the total pressure to be constant (CP case) at all depths of
the cloud. In the CP assumption, the density is defined only at the illuminated surface
of the cloud, which due to radiation compression gets stratified across the depths of
the cloud. In the CP models, the extent of density stratification does not only depend

© Springer Nature Switzerland AG 2019   35
T. P. Adhikari, *Photoionization Modelling as a Density Diagnostic
of Line Emitting/Absorbing Regions in Active Galactic Nuclei*,
Springer Theses, https://doi.org/10.1007/978-3-030-22737-1_3

on the strength of the illuminating radiation field but also on the hydrogen gas density defined at the surface of the gas cloud, i.e. on the surface gas pressure.

Moreover, it has been shown in many studies that if the ionizing radiation shape is dominated by the UV/soft X-ray photons, the ionization and thermal structure do depend on the hydrogen density used in the photoionization modelling [5, 6]. This is because of the reason that the dominant gas heating/cooling mechanisms behave differently in the gas clouds of different hydrogen densities. The difference in gas heating and cooling behaviour becomes most significant when the comparison is done between the gas clouds with large variation in gas density. In this Chapter, I investigated the influence of gas density on the overall structure of CD cloud. In addition, a detailed comparison between CD and CP individual cloud is done, and all possible solutions of hydrostatic equilibrium are considered. Furthermore, stability curves obtained by CLOUDY and TITAN models are compared. The influence of gas density on the single CP cloud is analyzed at the end of the Chapter.

3.1   Mrk 509 and Its SED

To make this Chapter clear I use only one shape of the SED as an input to the photoionization computations. Mrk 509 is a Sy1.5 galaxy, also considered to be the closest QSO/Sy1 hybrids and harbours a supermassive black hole of mass 1.4×10^8 $M_\odot$ [7]. It is one of the best studied AGN in the local universe with redshift of 0.034397 [8] and exceptionally high luminosity $L(1 - 1000 \text{ Ryd}) = 3.2 \times 10^{45}$ erg s^{-1}. The X-ray outflow in this source has been extensively studied using the 600 ks RGS on board of *XMM-Newton* X-ray telescope [9, 10]. The good spectral resolution of RGS allowed to identify few tens of absorption lines from highly ionized metals and hence the determination of ionic column densities. From the derived ionic column densities Detmers et al. [9] computed the equivalent hydrogen column densities and constructed the AMD. Additionally, by using the variability method, which is based on the assumption that the observed changes are due to the changes in the ionization states, Kaastra et al. [10] estimated the upper limits on the location of different components of the warm absorber in the range of 5–400 pc.

Kaastra et al. [11] published up to now the best broad band spectral shape of Mrk 509, which was obtained by combining the data from their multiwavelength observations campaign using various instruments of *XMM Newton*: RGS, EPIC (pn, MOS), and OM. Moreover, their campaign also consists of simultaneous data from *INTEGRAL* in hard X-rays as well as *Swift*: XRT and UVOT. The final SED was constructed including the infra-red points from the *IRAS* and *Spitzer* data, though it is not yet well understood how much infra-red radiation is covered by outflow. This is because the outflowing X-ray wind is most probably closer to the central black hole than the dusty torus from where the IR radiation is emitted.

With the authors permission, the SED of Mrk 509 is utilized for modelling the ionized absorber in this source. One of the reason for choosing this source in our study is the availability of well constrained SED based on the data obtained from the observations. The coverage of a wide range of wavelength band is very essential for obtaining the ionization balance required for photoionization modelling. Additionally, Mrk 509 is among the best studied AGN for which many observational properties of the outflow are available that can be directly compared with results obtained from the numerical simulations. I present the input SED of Mrk 509 in Fig. 3.1 where the observed spectral points are shown by red circles and the linearly interpolated SED based on these points is shown by the black solid line and it is used in both codes TITAN and CLOUDY for all the photoionization computations done in this Chapter.

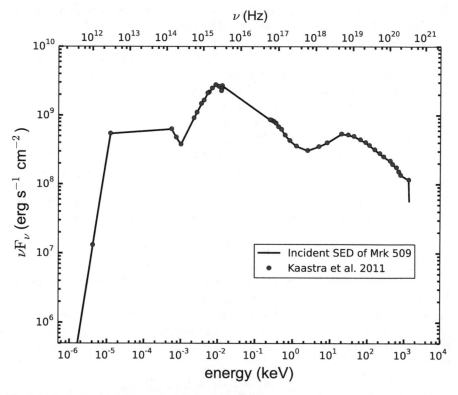

Fig. 3.1 Spectral energy distribution of Mrk 509. The black line corresponds to the incident spectrum which is the input to the photoionization codes TITAN and CLOUDY. The red circles are the points adapted from Kaastra et al. [11] and normalised to the incident flux. The figure is reproduced from Adhikari et al. [12]. ©AAS. Reproduced with permission

3.2   Single CD Cloud

In order to investigate how the irradiated cloud properties are influenced by the gas density values, I compute the CD CLOUDY models for different values of n_H. At a given cloud location r, other model parameters: the chemical composition and total column density N_H are set equal. CLOUDY self consistently computes the U from the source luminosity and the cloud radius through Eq. 2.12. In this Section, I use the luminosity of the source to be 10^{45} erg s^{-1}. All such CD clouds are subjected to the radiation field of Mrk 509 shown in the Fig. 3.1. The resulting ionized (H$^+$) and neutral (H^0) ion densities plotted against the total column density across the depth of the cloud are presented in the Fig. 3.2 for different n_H.

The top row of the Fig. 3.2 show the relative H-ion densities obtained for the clouds located at the same distance $r = 0.01$ pc but with two values of $n_H = 10^{13.0}$ (left panel) and $10^{10.9}$ cm^{-3} (right panel) respectively. I found that the column density at which the H$^+$ layer forms for the model with $n_H = 10^{13.0}$ cm^{-3} is smaller than the depth of the H$^+$ layer for the model with $n_H = 10^{10.9}$ cm^{-3}. The similar results are seen if we compare the left and right panels of the middle and bottom row of the figure where the comparison between other density values i.e., $n_H = 10^{11.5}$ cm^{-3} and $10^{9.4}$ cm^{-3} ($r = 0.1$ pc), and $10^{8.5}$ cm^{-3} and $10^{6.4}$ cm^{-3} ($r = 10.0$ pc) is done. This clearly demonstrates that the thickness of the ionization front depends on the gas density for three different cloud locations.

The dependence of the size of the ionization front on the density influences the line emission properties of the gas. This is related with the fact that the lines are formed inside the ionization front. Ionization front is a boundary that separates the fully ionized and neutral region of the gas cloud. To understand this physically, let us consider the example of Hβ line and assume simple ground-state hydrogen photoionization. Then, the ionization balance equation [14] can be written as

$$n_p\, n_e\, l\, \alpha_B(T) = \phi_H, \tag{3.1}$$

where n_p and n_e are the proton and electron densities in the H$^+$ layer of the cloud, l [cm] is the thickness of the H$^+$ layer, and $\alpha_B(T)$ [cm^3 s^{-1}] is the Case B recombination coefficient. The physical interpretation is that the flux of hydrogen-ionizing photons incident on the cloud, ϕ_H [cm^{-2} s^{-1}], equals the number of hydrogen recombinations that occur over the thickness l. Then the hydrogen column density across H$^+$ layer is

$$N_{H^+} \equiv n_p l = \frac{\phi_H}{n_e \alpha_B(T)} = U c\, \alpha_B(T)^{-1}. \tag{3.2}$$

The gas column density N_{H^+}, and the line and continuum optical depths, all depend on the ionization parameter. At very high values of U, whole cloud is ionized, and the line emission comes from the whole volume, in this case limited by the fixed adopted total hydrogen column density, N_H. As the cloud density increases, the ionization parameter decreases, ionization front forms closer to the cloud surface.

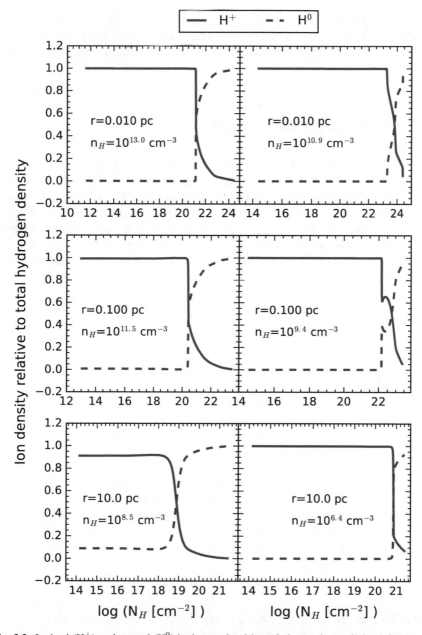

Fig. 3.2 Ionized (H$^+$) and neutral (H^0) hydrogen densities relative to the total cloud density n_{H} as a function of the total column density of the single cloud illuminated by Mrk 509 SED. The left and right panels in the top, middle and bottom rows show the ion density comparisons for n_{H}: $10^{13.0}$ and $10^{10.9}$ ($r = 0.01$ pc), $10^{11.5}$ and $10^{9.4}$ ($r = 0.10$ pc), and, $10^{8.5}$ and $10^{6.4}$ cm^{-3} ($r = 10.0$ pc) respectively. The figure is reproduced from Adhikari et al. [13]. ©AAS. Reproduced with permission

This is clearly demonstrated in Fig. 3.2, which shows that the H^+–H^0 ionization front moves towards smaller cloud thickness with increasing number density of the cloud.

I applied the density dependence of ionization front and investigated its consequences in the resulting line emissivity profiles in the Chap. 5.

3.3 CP Versus CD Single Clouds

In this Section, I present the basic theoretical considerations of a gas cloud in the hydrostatic equilibrium. Let us consider a single cloud in a spherically symmetric gravitational field, located at a radial distance r from the SMBH. The main assumption is that locally the cloud thickness is negligible in comparison to the distance which is equivalent to a locally plane parallel approximation of the cloud geometry. The cloud is illuminated by a radiation flux F_0 at the illuminated face z_0. The condition of hydrostatic equilibrium, for black hole mass M_{BH}, with optical depth as a variable $d\tau = \kappa\rho dz$, is:

$$\frac{dP_{gas}}{d\tau} = -\frac{1}{\kappa}\left(\frac{GM_{BH}}{r^2} - \Omega^2 r\right) - \frac{dP_{rad}}{d\tau}, \tag{3.3}$$

where as usual κ is mean opacity coefficient, G—gravitational constant and Ω is the gas angular velocity. Expressing the radiation pressure gradient as a first order solution of radiative transfer equation: $dP_{rad}/d\tau = -(F_0/c)\,e^{-\tau}$, and integrating of hydrostatic balance from 0 to τ, we obtain:

$$P_{gas}(\tau) = P_{gas}(0) - \frac{1}{\kappa}\left(\frac{GM_{BH}}{r^2} - \Omega^2 r\right)\tau - \frac{F_0}{c}e^{-\tau} + \frac{F_0}{c}. \tag{3.4}$$

At the illuminated face of the cloud, the total pressure has some constant initial value: $C_0 = P_{gas}(0) + F_0/c$. In addition, the gas pressure gradient at the outer cloud surface ($\tau = \tau_{max}$) should be zero, since the cloud is finite and the radiation pressure and centrifugal force should balance the gravitational force there i.e. $(GM_{BH}/r^2 - \Omega^2 r)/\kappa = (F_0/c)\,e^{-\tau_{max}}$. Adopting these conditions, the hydrostatic balance is:

$$P_{gas}(\tau) = C_0 - \frac{F_0}{c}(e^{-\tau} + \tau e^{-\tau_{max}}) \tag{3.5}$$

To put limits on the initial conditions of cloud pressure which fulfils the hydrostatic balance we express the ratio of gas pressure at two extreme cases: for $\tau = 0$ and $\tau = \tau_{max}$

$$\frac{P_{gas}(\tau_{max})}{P_{gas}(0)} = \frac{C_0 - \frac{F_0}{c}e^{-\tau_{max}}(1 + \tau_{max})}{C_0 - \frac{F_0}{c}}. \tag{3.6}$$

The above equation does not put any limit on the initial values of cloud pressure being in pressure equilibrium. We can consider special limits. Assuming $\tau_{max} \ll 1$, we get:

$$\frac{P_{gas}(\tau_{max})}{P_{gas}(0)} = 1; \quad \Rightarrow P_{gas} = \text{const}, \tag{3.7}$$

which is well known case of constant gas pressure cloud. It means that for optically thin clouds we do not have strong modification of the gas pressure by the radiation pressure. Second special case occurs for $\tau_{max} \gg 1$, and we get

$$\frac{P_{gas}(\tau_{max})}{P_{gas}(0)} = \frac{C_0}{C_0 - \frac{F_0}{c}} = 1 + \frac{P_{rad}(0)}{P_{gas}(0)}. \tag{3.8}$$

In this case the density gradient inside the cloud depends on the adopted value of gas to radiation pressure at the illuminated face of cloud. When $P_{rad}(0) \ll P_{gas}(0)$ again we get condition of cloud being under constant gas pressure. But when $P_{rad}(0) \geq P_{gas}(0)$, the compression of cloud by radiation pressure is always present and increases with increasing value of this ratio.

Thus the requirement that radiation pressure should be much larger from the gas pressure, made by Baskin et al. [4] for the RPC model, is only a special case among solutions for CP cloud, where the compression is the strongest. Physically, it is always the case for the WAs in AGN modelled by Różańska et al. [5]. Nevertheless, many other solutions of clouds being under CP are possible from small compression equivalent to constant gas pressure model up to strong RPC considered by Baskin et al. [4] in case of BLR.

To illustrate physical differences between single CP and CD clouds, I present in Fig. 3.3 the structure of a single cloud as a result of photoionization calculations with CLOUDY code. Our cloud is located at 1 pc from SMBH and illuminated by SED of Mrk 509. The assumed density on the cloud surface is $n_H = 10^{10}$ cm^{-3}. The difference between assumption of CD (dashed blue line) and CP (solid red line) is noticeable. The density structure for CP cloud is not constant, even when compression is very weak, since radiation pressure is gradually absorbed with cloud optical depth.

Lower gas pressure at illuminated cloud surface can relatively increase the compression by radiation pressure. Physically this condition can be achieved for the lower density. In Fig. 3.4, I present the same single cloud comparison for density 3.16×10^5 cm^{-3}. For such cases we are in the limit where $P_{rad}(0) \geq P_{gas}(0)$ and compression is clearly visible as a density and gas pressure rise up with cloud thickness.

In Fig. 3.5, I show the position of the H-ionization front in case of CP and CD models. For the low density 3.16×10^5 cm^{-3}, it is found that the compression by radiation pressure produces H$^+$ at depths $\sim$3 orders of magnitude lower than in the case of CD models. However this difference is very insignificant when the density of the gas clouds reaches to 10^{10} cm^{-3}. The consequence of the density effects in CD and CP clouds in the line emission is shown in the Fig. 3.6. From the figure it

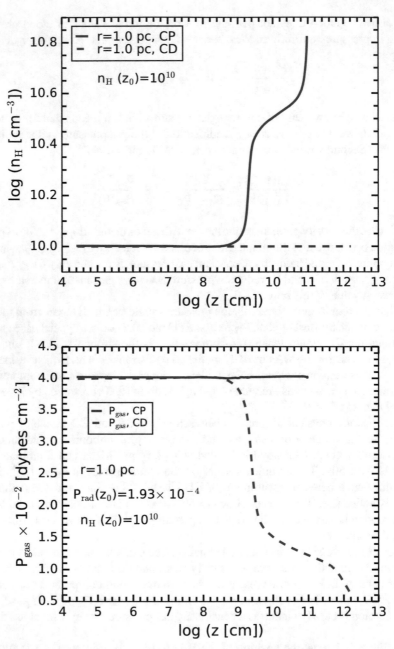

Fig. 3.3 Upper panel shows the stratification in density when the gas clouds defined by the initial gas density $n_H(z_0) = 10^{10}$ cm^{-3} is illuminated by the SED of Mrk 509. The lower panel exhibits the pressure structure for the same set up as in upper panel. Solid red line shows the case for CP whereas the dashed blue line is for CD case. The figure is reproduced from Adhikari et al. [15]. ©AAS. Reproduced with permission

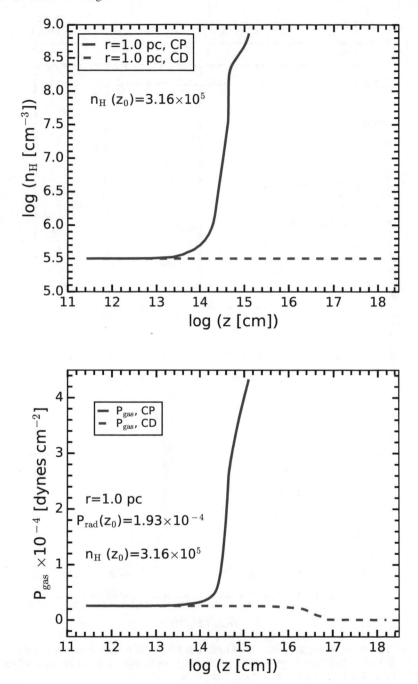

Fig. 3.4 Same as that of Fig. 3.3 but for the gas density $n_H (z_0) = 3.16 \times 10^5$ cm^{-3}. The figure is reproduced from Adhikari et al. [13]. ©AAS. Reproduced with permission

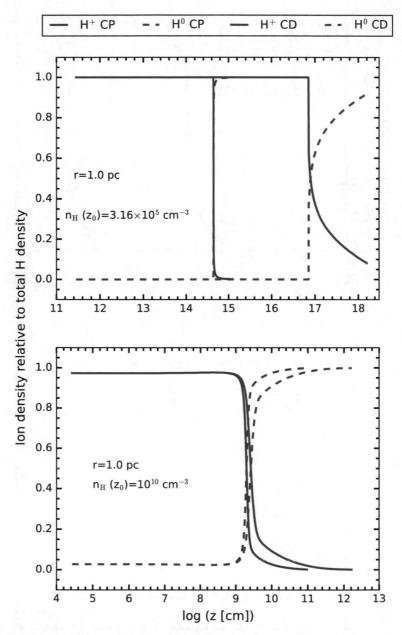

Fig. 3.5 Plot of the Hydrogen ion densities relative to the total H density as function of the cloud depths. The upper and lower panel represent the case for the density at the illuminated surface of the slab equaled to 3.16×10^5 and 10^{10} cm^{-3} respectively

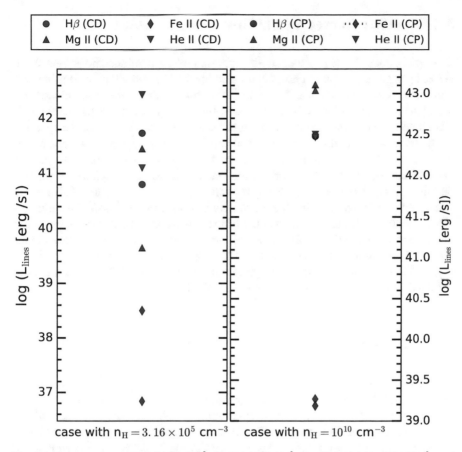

Fig. 3.6 A comparison of Hβ λ4861.36 Å, He II λ1640.00 Å, Fe IIλ (4434.00–4684.00) Å and Mg II λ2798.0 Å line luminosities under CD (blue) and CP (red) assumptions. The left and right panel represent the case for the density at the illuminated surface of the slab equaled to 3.16×10^5 and 10^{10} cm^{-3} respectively

is evident that when the gas density is low enough to satisfy the RPC condition, the line luminosity for Hβ, Mg II, Fe II and He II in CP case is higher by $\sim$1.5 orders of magnitude than in CD case. However in case of high density gas, the difference in line luminosity between CD and CP case is negligible. The detailed comparison of the CP and CD clouds in the context of the emission line regions in AGN is done in the Chap. 5

3.4 Stability Curves for CLOUDY and TITAN Models

The significance of studying the stability curve is already mentioned in the Sect. 2.3.1. Here, I present the stability curves computed for the CD and CP models using both numerical codes CLOUDY and TITAN, in the context of warm absorbing clouds. For the plane parallel geometry (see Sect. 2.2.4 for details) of the clouds, the multiple CD models are required to construct the full stability curve. In case of the spherical geometry achieved under certain physical conditions even a single CD cloud can go through a wide range of ionizations and the full stability curve is possible since in CLOUDY code geometrical dilution of illuminated radiation is taken into account (see Section below). However, the CP assumptions with $P_{rad} >> P_{gas}$ demands the self consistent stratification in the thermal and ionization structure inside the single gas cloud which enables us to construct the full stability curve. In all the computations presented in this Section, Solar values of the chemical compositions shown in the Table 2.1 are used. Additionally, the SED shape of the Mrk 509 (shown in the Fig. 3.1) is normalized to the hydrogen ionizing luminosity 3.2×10^{45}, the value obtained from the observations by Kaastra et al. [11].

3.4.1 CD Models

I present the comparison of the stability curves computed using different assumptions in the Fig. 3.7. The stability curve obtained from the multiple cloud models of the WA computed with CLOUDY is shown by cyan triangles whereas that obtained with the TITAN is represented by red circles. The parameters employed for the set of models in our computations with both codes are listed in the Table 3.1. For each case the distance of the illuminated side of the cloud from the source of ionizing radiation is taken as $\log(r/cm) = 17.25$.

When the geometry is spherical and the shells of gas clouds are extended, the radiation pressure falls faster as we go deeper into the cloud i.e., $P_{rad} \propto 1/r^2$ because of the geometrical dilution and the absorption by ionized gas. This case is achieved in CLOUDY when a very less dense cloud is used, in this case the density being 1 cm^{-3}. The geometrical dilution causes the decrease in ionization parameter and hence the temperature of the gas decreases even for the constant density at each cloud zones.

Table 3.1 CD cloud parameters. When the range of densities is shown, it means the grid of clouds were calculated. The inner cloud radius in all cases is $\log(r/cm) = 17.25$

Codes used	n_H (cm^{-3})	ξ_0 (ergs cm s^{-1})	Geometry
TITAN	10^2 to 10^{12}	as $L/n_H r^2$	Plane parallel
CLOUDY	10^2 to 10^{12}	as $L/n_H r^2$	Plane parallel
CLOUDY	1	10^{11}	Spherical

So, in principle it is possible to construct a full stability curve with a single constant density spherically symmetric and geometrically extended model of an ionized gas. The stability curve obtained by this procedure is depicted by a magenta solid line in the Fig. 3.7.

I found that all the three stability curves agree in warm absorber regime i.e., down to the temperature $\sim 3 \times 10^4$ K and the disagreement appears below this value. This may be because of the differences between the two codes; most probably in the atomic database used. However, for the purpose of studying the WA with CD assumptions, both codes work perfect.

3.4.2   CP Models

The stability curves computed with TITAN and CLOUDY assuming that the WA is a single gas cloud under CP equilibrium are shown in Fig. 3.8. Both models are

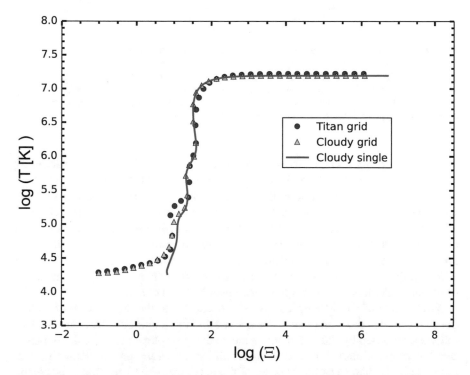

Fig. 3.7 Stability curves for CD plane-parallel clouds obtained with the photoionization codes TITAN (red circles), and CLOUDY (cyan triangles). Additionally, CD single cloud with $n_H = 1\,\mathrm{cm}^{-3}$ in spherical geometry computed with CLOUDY is presented by solid magenta line. This plot is reproduced from Adhikari et al. [12]. ©AAS. Reproduced with permission

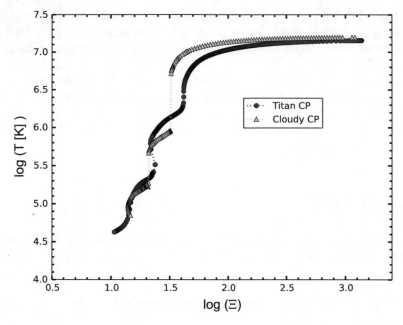

Fig. 3.8 Stability curve obtained from CLOUDY (red circles) and TITAN (cyan triangles). Both curves are produced with the number density $n_H = 10^6$ cm^{-3} and ionization parameter $\xi_0 = 4.45 \times 10^5$ erg cm s^{-1} at the irradiated surface of the cloud. The inner radius to the cloud is set to log(r/cm) = 16.75, to ensure that illuminating luminosity is equal to the observed one. The total column density (N_H) of the slab in both calculation is set to 1.5×10^{23} cm^{-2}. The figure is reproduced from Adhikari et al. [12]. ©AAS. Reproduced with permission

computed with the parameters: density $n_H = 10^6$ cm^{-3} and ionization parameter $\xi_0 = 4.45 \times 10^5$ erg cm s^{-1} at the irradiated surface of the cloud.

The TITAN points constituting the stability curve are marked by the red circles while the CLOUDY points are shown by the cyan upper trianlges. The overall trend of the stability is similar except in the regions of thermal instability where there are considerable differences visible. In particular, the thin unstable regions of the curve are not well computed by the CLOUDY code i.e., the points in the region of the curve with negative slope are missing. However, the TITAN code niccly depicts the layers of the thermally unstable zones in the computed CP gas cloud.

While some of the differences may also occur due to the different set of atomic databases used in the numerical codes, I argue that this significant difference seen in our work is caused by the difference in the methods employed to solve the radiative transfer computations in both codes. In our opinion, TITAN computes radiative transfer more correctly than CLOUDY because of the accuracy of ALI method over escape probability method when an optically thick gas cloud is under consideration and the differences are mostly noticed while computing the photoionization models for the gas clouds satisfying the condition $P_{rad} >> P_{rad}$ discussed in the Sect. 3.3. For this reason, I used TITAN to study AMD of AGN outflow in Chap. 4.

3.5 The Effect of Density on the CP Models of WA

The influence of the gas density on the various properties of the WA structure are computed using the TITAN models under CP assumption. In this Section, I consider the CP clouds for various values of gas densities: $\log(n_H [\text{cm}^{-3}]) = 5, 6, 8$ and 9 at the cloud surface, each of them being illuminated by the radiation field of Mrk 509. For all the clouds, the ionization parameter at the surface is set to $\xi_0 = 4.3 \times 10^4$ erg cm s^{-1} and the Solar values of the chemical compositions are used.

In Fig. 3.9, I present the plot of heating-cooling mechanisms that operate inside the illuminated gas clouds under CP condition. Here, I consider the most dominant processes of the heating and cooling i.e., Compton and free-free contributions (see Sect. 2.1 for definition). I found a clear trend of increase in free-free heating for increase in the gas density. The free-free heating at the cloud surface increases by

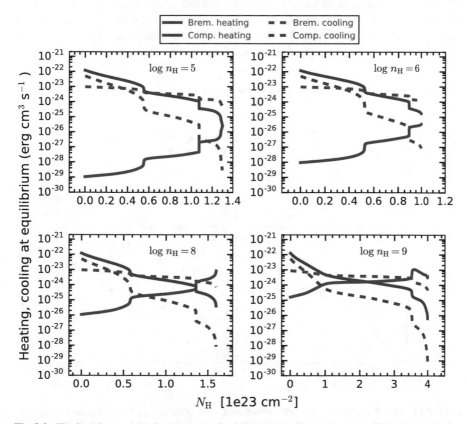

Fig. 3.9 The free-free and the Compton contribution to the heating and cooling of the gas of various values of $\log n_H$ ([cm^{-3}]) $= 5, 6, 8$ and 9 are shown for CP clouds. A clear trend of increase in free-free heating is seen on increasing the gas density. The models are computed with the value of $\xi_0 = 4.3 \times 10^4$ erg cm s^{-1}

~4 orders of magnitude while moving from the density values 10^5 to 10^9. Deep inside the densest cloud considered in this case (lower right panel of Fig. 3.9), the free-free heating dominates over the Compton heating. Moreover, the two noticeable jumps in heating-cooling processes for $n_H \leq 10^8$ cm^{-3} reduces to a single jump when the density is increased to 10^9 cm^{-3}. These results are consistent with the previous studies done by Różańska et al. [5], Chakravorty et al. [6] where they showed that photoionization models depend on the gas density if the incident SED is dominated by the soft X-ray/UV photons.

The thermal and ionization structures: the pressure, the gas density, the ionization parameter and the temperature along the layers of an absorbing cloud as predicted by TITAN are shown in the Fig. 3.10 for two representative gas density values. The solid lines show the WA properties for the density $n_H = 10^8$ cm^{-3} at the surface of the slab, whereas the dashed lines show the WA properties for the density $n_H = 10^9$ cm^{-3}. The lower left panel of the figure shows stratification in the gas density when the absorbing cloud is assumed to be in CP. It is clear that the stratification happens faster for the less dense cloud for the reason that $P_{\text{rad}}/P_{\text{gas}}$ is much larger than in the case of high density cloud. Consequently, the less dense clouds become cooler much faster then the high density clouds (upper right panel of the figure). For $n_H = 10^8$ cm^{-3} model,

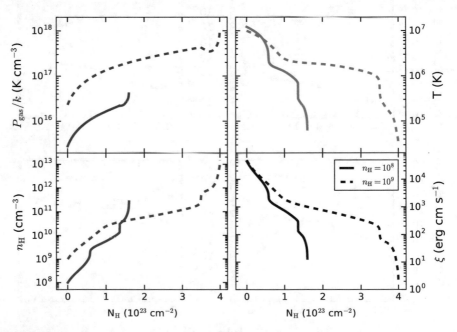

Fig. 3.10 The structure of pressure (upper left), temperature (upper right), density number (bottom left), and ionization parameter (bottom right), versus cloud column density for initial parameters $n_H = 10^8$ cm^{-3}, $\xi_0 = 4.3 \times 10^4$ erg cm s^{-1} (solid lines) and $n_H = 10^9$ cm^{-3}, $\xi_0 = 4.3 \times 10^4$ erg cm s^{-1} (dashed lines). In both cases, the CP models are shown for illuminating luminosity $L = 3.2 \times 10^{45}$ erg s^{-1} and Mrk 509 SED. This plot is reproduced from Adhikari et al. [12]. ©AAS. Reproduced with permission

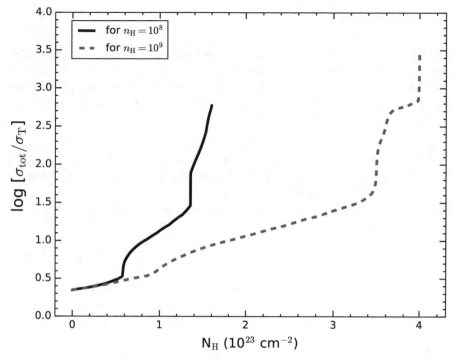

Fig. 3.11 Total cross section (σ_{tot}) relative to Thomson cross section (σ_T) as a function of cloud column density. The initial cloud parameters are the same as in Fig. 3.10. The figure is reproduced from Adhikari et al. [12]. ©AAS. Reproduced with permission

there are two noticeable temperatures drops between $T = (4 - 9) \times 10^5$ and $T = (2 - 5) \times 10^6$ K, shown by the green solid line in the figure. This clearly demonstrates that if the absorbing material is in pressure equilibrium, then the thermally unstable thin regions can be clearly visible. These drops correspond to the negative slope part of the thermal instability curves depicted in the Fig. 3.8.

I plotted the total cross section, σ_{tot}, relative to the Thomson scattering cross section, σ_T, as a function of the absorbing column density of the cloud for the two initial gas densities $n_H = 10^8$ (solid line) and 10^9 cm^{-3} (dashed line) in the Fig. 3.11. Two cases of sudden rise of the total cross section occur at the position corresponding to the fall of temperature shown in the Fig. 3.10 (upper right panel).

I note here that the thermally unstable zones are geometrically very thin and it is not well resolved by radiative transfer codes. Różańska [16] have shown that the correct temperature profiles in the thin, unstable transition zones can be recovered only after including the electron conduction which at the moment is not included in any of the radiative transfer codes. However, the location of the discontinuity is pretty well determined by solving the radiative transfer only, and its extension depends on the model parameters as was shown by Czerny et al. [17].

3.6   Conclusions

In this Chapter, I studied the influence of the gas density in the photoionization models computed with the SED shape of Mrk 509. Both the numerical codes CLOUDY and TITAN were used to derive the cloud properties under CD and CP assumptions. From the derived cloud properties and its dependence on the assumed gas densities, I reached to the following conclusions:

1. The study of CD single gas clouds revealed that the depth of the H-ionization front is smaller for the high density gas as compared to the less dense gas. This dependence has consequences in the line emission properties since line emission comes from the ionized part of the gas.
2. The radiation pressure confinement is only a special condition of a CP solution which is achieved for the condition $P_{rad} >> P_{gas}$. For the same SED, RPC conditions depend on gas density used in the computations.
3. For the density of the order of 10^5 cm^{-3}, and typical AGN luminosity, CP single cloud model differs from CD model. The luminosities of major emission lines Hβ, Mg II, Fe II and He II for CP cloud are 1.5 order of magnitude higher than for CD cloud. Therefore, CP models provide higher line luminosities and may be an alternative explanation for line emitting regions of lower densities, located close to NLR.
4. The comparison of the stability curves computed with the codes CLOUDY and TITAN demonstrates that both code agree well for the CD assumptions of the gas clouds. Nevertheless, TITAN offers better resolution of the very thin thermally unstable zones of the CP clouds due to the more accurate treatment of the radiative transfer solution.
5. For the Mrk 509 like SED, the heating-cooling mechanism significantly depends on the gas density. Deep inside the dense gas cloud, the free-free process becomes the dominant gas heating mechanism.
6. There is a clear evidence that the CP cloud structure depends on density. For Mrk 509 SED shape, lower gas density ($n_H < 10^8$ cm^{-3}) cloud exhibits two unstable zones while in the clouds with $n_H = 10^9$ cm^{-3}, one unstable zone is seen.

I applied these conclusions of density dependence of photoionization models and investigated its consequences in the modelling of: the AMD (Chap. 4), and the line emitting regions (Chap. 5) using various SED shapes of AGN radiation.

References

1. Różańska A, Czerny B, Siemiginowska A, Dumont A-M, Kawaguchi T (2004) ApJ 600:96
2. Różańska A, Goosmann R, Dumont A-M, Czerny B (2006) A&A 452:1
3. Stern J, Laor A, Baskin A (2014) MNRAS 438:901
4. Baskin A, Laor A, Stern J (2014) MNRAS 438:604
5. Różańska A, Kowalska I, Gonçalves AC (2008) A&A 487:895
6. Chakravorty S, Kembhavi AK, Elvis M, Ferland G (2009) MNRAS 393:83
7. Peterson BM et al (2004) ApJ 613:682

8. Huchra J, Latham DW, da Costa LN, Pellegrini PS, Willmer CNA (1993) AJ 105:1637
9. Detmers RG et al (2011) A&A 534:A38
10. Kaastra JS et al (2012) A&A 539:A117
11. Kaastra JS et al (2011) aap, 534, A36
12. Adhikari TP, Różańska A, Sobolewska M, Czerny B (2015) ApJ 815:83
13. Adhikari TP, Różańska A, Czerny B, Hryniewicz K, Ferland GJ (2016a) ApJ 831:68
14. Osterbrock DE, Ferland GJ (2006) Astrophysics of gaseous nebulae and active galactic nuclei
15. Adhikari TP, Hryniewicz K, Różańska A, Czerny B, Ferland GJ (2018a) ApJ 856:78
16. Różańska A (1999) MNRAS 308:751
17. Czerny B, Chevallier L, Gonçalves AC, Różańska A, Dumont A-M (2009) A&A 499:349

Chapter 4
Absorption Measure Distribution (AMD) in AGNs

Abstract In this Chapter, I present the AMD models computed with the numerical code TITAN under the assumption that the X-ray absorber in AGN is in the total pressure equilibrium. By simulating such ionized gas cloud, the distribution of ionization parameter across the depth of the gas cloud is calculated. For each computed model, the AMD is derived taking into account the stratification of the ionization parameter. In order to investigate the effect of different illuminating radiation, I employed various types of AGN SEDs in my computations. Moreover, the dependence of AMD models on various gas cloud parameters is studied. Finally, the computed models are then compared with the currently available observationally derived AMDs for various sources.

In Chap. 3, I showed that the density parameter n_H, defined at the surface of the photoionized cloud is important in shaping the overall structure of the gas, both in emission and absorption. Moreover I demonstrated that, the ALI method of radiative transfer computations in TITAN allows us to resolve the thin layers of strong temperature and density gradients better, developed in the photoionized gas cloud. In this Chapter, I present the AMD models computed with the numerical code TITAN under the assumption that the X-ray absorber in AGN is in the total pressure equilibrium.

For obtaining AMDs from photoionization simulations, I considered a CP single gas cloud defined by the parameters: density n_H and ionization parameter ξ_0 both assumed at the surface, and the total column density N_H (see Chap. 2 for the parameters description). Such a cloud, on illumination by the AGN radiation field is consistently stratified into various layers with gradients of density and temperature. Since the normalization of the incident radiation flux is determined by the $\xi\,n_H$ product in TITAN code, the information about the source luminosity and the distance to the cloud is not required as an input parameter. In all cases considered in this Chapter, the shape of the radiation flux which enters an absorber is given by points, and the linear interpolation is done in order to fully define the incident continuum for the TITAN code.

By simulating such ionized gas cloud, the distribution of ionization parameter across the depth of the gas cloud is calculated using the Eq. 2.10. The structure of both pressures (gas and radiation) with column density of a layer, is an output of the

© Springer Nature Switzerland AG 2019

T. P. Adhikari, *Photoionization Modelling as a Density Diagnostic of Line Emitting/Absorbing Regions in Active Galactic Nuclei*, Springer Theses, https://doi.org/10.1007/978-3-030-22737-1_4

TITAN code. For each computed model, the AMD is derived using the relation given in the Eq. 2.13, taking into account the stratification of the ionization parameter ξ.

In order to investigate the effect of different illuminating radiation, I employed various types of AGN SEDs in my computations. Moreover, the dependence of AMD models on various gas cloud parameters: n_H, ξ_0 and N_H is studied. Finally, the computed models are then compared with the currently available observationally derived AMDs for various sources.

4.1　AMD for Mrk 509

In this Section, I consider the case of the Sy1.5 galaxy, Mrk 509, for which the detail observational studies about the spectral and the absorption properties are available as summarized in Sect. 3.1. All the models are considered to be in CP and the Solar chemical composition is employed in all computations.

In Fig. 4.1, I show the influence of the gas density on the AMD for Mrk 509 TITAN models for varying values of densities: $n_H = 10^5$, 10^6, 10^8 and 10^9 cm^{-3} assumed

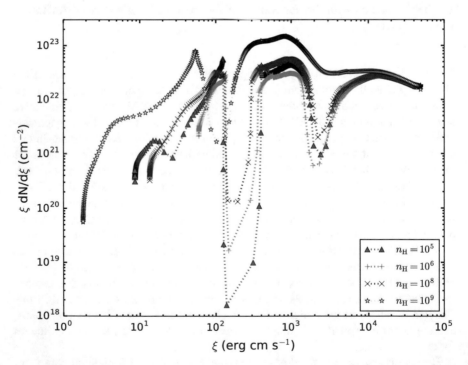

Fig. 4.1 AMD structure for the same initial ionization parameter $\xi_0 = 4.3 \times 10^4$ erg cm s^{-1} and different values of n_H at the illuminated side of the cloud marked in right bottom corner. The figure is reproduced from Adhikari et al. [1]. ©AAS. Reproduced with permission

at the illuminated face of the clouds. Two prominent discontinuities in the AMD, located between the log $\xi \sim 2$–3 and 3–4 are seen in the modelled AMD, which are most probably caused by the thermal instability developed inside the cloud. The depths of the cloud at which the dips exist in the AMD remain the same for the models with lower values of gas density i.e., n_H upto 10^8 cm^{-3}. As the density increases further, the dip in AMD changes. This is an important result as it shows that the AMD analysis can be used as a potential density diagnostic of the absorbing plasma in AGNs.

The changing behaviour of AMD dip with density is connected with the fact that the photoionization models of the cloud with gas density similar to that of a BLR gas are not degenerate when the illuminating spectral radiation of the source is dominated by the strong component of optical/UV part [2]. And this is the case for Mrk 509. The physical reason for this is that, in case of the hard spectrum, all radiative processes (Comptonization, line heating/cooling) are linear with density and the solution of photoionized numerical calculations is determined by the ionization parameter, independently from the local density. However, for the soft incident spectra the free-free process plays an important role and the quadratic dependence on the density breaks the degeneracy and introduces the dependence on the density

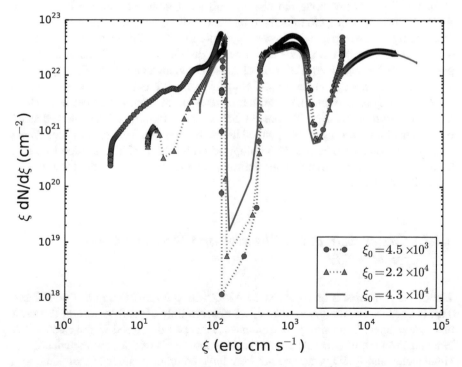

Fig. 4.2 AMD structure for the same initial number density $n_H = 10^6$ cm^{-3} and different values of ionization parameters at the illuminated side of CP clouds marked in right bottom corner. The figure is reproduced from Adhikari et al. [1]. ©AAS. Reproduced with permission

[2]. In Sect. 3.5, I have shown that the increase in the free-free heating on moving from the low to high gas density occurs in the case of Mrk 509 SED and ultimately becomes the dominant mechanism deep inside the dense cloud (see Fig. 3.9).

When the density of an ionized gas is comparable to that of NLR, the AMD dips remain prominent near the same range of ionization states. Moving towards the density regime comparable to BLR density, i.e., at $n_H = 10^9 \, cm^{-3}$, only a single prominent dip survives. The overall normalization of the AMD does not vary significantly for densities upto $10^8 \, cm^{-3}$, but the depths of the corresponding dips are different. Dips are less prominent for the higher densities assumed at the surface of the absorber.

The dependence of AMD structure on the ionization parameter ξ_0 defined at the irradiated surface of the absorber is shown in the Fig. 4.2. It is found that the AMD dips occur around the same ionization range independent of the initial ionization parameter ξ_0, although the depth of the dips around $\log \xi = 2$–3 slightly differs among these models.

I note here, that there is no difference in AMD from the clouds of the same density affected by various luminosities if we assume the same ionization parameter. This is obvious from Eq. 2.9 due to the fact that for the same product of $\xi \, n_H$, the same energy flux L_{ion}/r^2 is illuminating the absorber.

In Fig. 4.3, I demonstrate the effect of different total column densities N_H on AMD for the absorbing cloud defined by the density $n_H = 10^8 \, cm^{-3}$. As it is seen from the figure, absorbers with lower N_H do not give two dips in AMD. This is due to the fact that the computations with low N_H do not provide sufficient material to pass through the second zone of thermal instability at lower ξ values. For $N_H > 10^{22}$ cm^{-2}, the AMD always has two dips. From the figure, it is evident that when the total column density is substantial, the further increase of this parameter does not change the distribution of the dips. This result demonstrates that the total column density only determines the range of ξ spanned in the distribution. This result is consistent with the CLOUDY models of AMD computed under the RPC assumption by Stern et al. [3], where the authors found that the value of N_H only affects the range of ionization structure.

4.1.1   Comparison of the Observed and Modelled AMD for Mrk 509

The direct comparison of an observed AMD with the modelled one is quite hard. This is because for the given source, we observe only several to several dozen of absorption lines. Each line puts a point on AMD as described in the Sect. 2.3.2. Note that, in case of computed models, a few thousands of lines are included in the simulations, and AMD is derived not from ionic column densities of individual ions but from overall cloud structure.

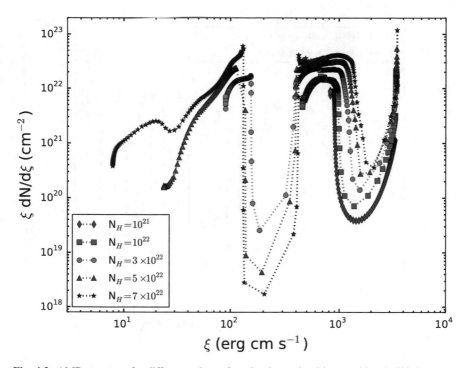

Fig. 4.3 AMD structure for different values of total column densities considered. This is a case with $n_H = 10^8$ cm^{-3} and $\xi_0 = 3.4 \times 10^3$ erg cm s^{-1}. Absorbers with low column densities, less then 10^{22} cm^{-2} do not have enough matter to pass through two dips in the AMD distribution. Those cases are indicated by blue diamonds and magenta squares. The figure is reproduced from Adhikari et al. [1]. ©AAS. Reproduced with permission

In Fig. 4.4, I show a comparison between the observed AMD (given by black histogram) for Mrk 509 [4] and the best TITAN model (red triangles). Our best fit model consists of a gas cloud under constant total pressure defined with the parameters: $n_H = 10^8$ cm^{-3}, $\xi_0 = 3.4 \times 10^3$ erg cm s^{-1}. The lower panel of the Fig. 4.4 shows the absolute values of AMD normalization whereas the AMD model scaled to the observed level is presented in the upper panel.

It is evident from the upper panel of the Fig. 4.4, that our model agrees very well with observed positions of the AMD dips. The discontinuities in AMD are in the range of the ionization parameter, $\log \xi = 2 - 3$ and $\log \xi = 3 - 4$. These ranges of ξ correspond to the temperature intervals $T = (4 - 9) \times 10^5$ and $T = (2 - 5) \times 10^6$ K, which are present in the temperature structure shown in Fig. 3.10 by the green solid line. This is a proof that the absorbing material should be in pressure equilibrium and the thin thermally unstable regions can be clearly seen as a decrease of the column density in the AMD. This comparison demonstrates that the warm absorber in Mrk 509 is the continuous cloud under CP. This also provides a strong constraint that gas density of the absorber in Mrk 509 is of the order of 10^8 cm^{-3} at the illuminated cloud surface. The two dips resulting from the gas irradiation are clearly

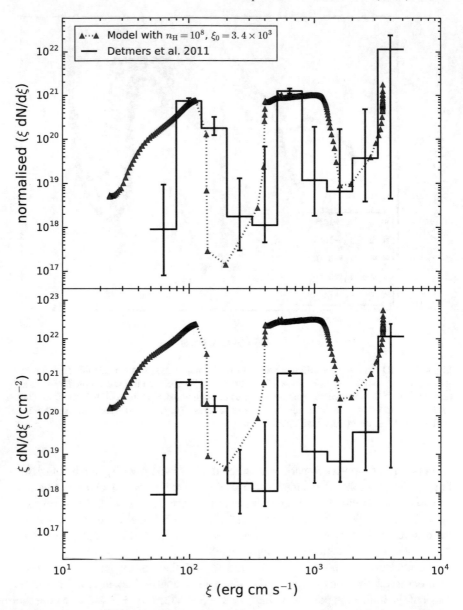

Fig. 4.4 The comparison of the simulated AMD structure—red triangles, with the observed one—black histogram. Data and error bars are taken from the paper by Detmers et al. [4]. The best model which roughly matches the data is computed for $n_{\rm H} = 10^8$ cm^{-3} and $\xi_0 = 3.4 \times 10^3$ erg cm s^{-1}. Bottom panel presents data with absolute normalization, while in the upper panel the modelled AMD is scaled down by a factor $\sim$30 to match the observed points on the stable branch. The best fit model represents well the depth of the two dips. The figure is reproduced from Adhikari et al. [1]. ©AAS. Reproduced with permission

seen in Fig. 3.10, where the various structures of CP cloud: gas pressure, temperature, density and the ionization parameter ξ as the function of column density across the depth are presented. Those AMD dips are directly connected with sudden rise of the total cross section σ_{tot} shown in Fig. 3.11, which occurs at the location consistent with the temperature drops.

As it is seen in the lower panel of Fig. 4.4, the observed AMD normalization of Mrk 509 corresponding to the ionization parameter $\xi = 8 \times 10^2 \, erg \, cm \, s^{-1}$ is $\sim 10^{21} \, cm^{-2}$ [4], less by a factor of ~ 30 than the AMD normalization from our best CP model with TITAN ($\sim 3 \times 10^{22} \, cm^{-2}$).

The difference in AMD normalization may be related to the fact that there are some considerable differences between the underlying assumptions of deriving the AMD from observations and from modelling. The modelled column densities presented in the figure by red triangles are more realistic column density, computed with radiation transmitted through all zones in the sequence, instead of transmission through a collection of independent zones, as used in observational data modelling by Detmers et al. [4] expressed in the Eq. 2.14. To check this hypothesis, the whole derivation of AMD from observations should be made, which is out of the scope of this thesis.

From the best TITAN model which matches observations for Mrk 509, I can derive the location of the warm absorber when the source luminosity is known. The luminosity measurements may be done by different methods, but the simplest way is to integrate broad-band spectrum available for this source. Taking integrated source luminosity to be 6.62×10^{45} erg s^{-1}, the wind location for the best fit model with $n_H = 10^8 \, cm^{-3}$ and $\xi_0 = 3.4 \times 10^3$ erg cm s^{-1} can be estimated as: $r = 1.39 \times 10^{17}$ cm. This value is an order of magnitude larger than the position of BLR, which is reasonable value for ionized gas clouds to be stable. My result is lower by a factor of 10–100 from the upper limits for the location of 5 different ionization components found by Kaastra et al. [5] from the variability method. This may be due to the fact that all data are fitted by CD models, and the theoretical AMD derived in this thesis is done for CP warm absorber. Despite this discrepancy, I present the first estimation of the WA distance done with the use of AMD for Sy1.5 bright AGN.

4.1.2 Transmitted Spectrum

The spectrum transmitted through the best fit model of the Mrk 509 SED, is presented in the Fig. 4.5. The energy range from 0.1 to 10 keV, typically used in currently working X-ray missions, is displayed in panel (a) of the figure. Numerous absorption lines are visible in the spectrum when the radiation is passing through the ionized gas. The absorption lines produced due to ions of lower ionization are shown in the panels (b) and (c) of Fig. 4.5. The absorption due to the different charge states of the Fe atom, i.e., around the 6.4 keV energy are shown in the panel (d). The advantage of CP model is that high and low ionization lines are obtained from the single gas cloud, naturally compressed by the radiation pressure.

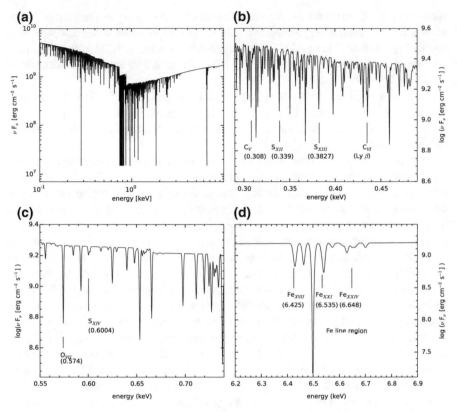

Fig. 4.5 Radiation spectrum transmitted through the slab of gas, for which modelled AMD agrees with Mrk 509 observed one demonstrated in Fig. 4.4. Panel **a** shows the spectrum spanning the energy band from 0.1 to 10 keV. Panel **b** shows the spectrum zoomed in the energy range 0.3–0.5 keV. Panel **c** shows the spectrum zoomed in the energy range 0.55–0.75 keV. In panel **d**, the spectrum is zoomed in the Fe-line absorption region around 6.4 keV. The figure is reproduced from Adhikari et al. [6] with the reprint permission from the Polish Astronomical Society

These spectral components resolved from the modelling demonstrate that the CP assumption of the absorbing clouds not only reproduce the absorption lines often seen in the UV/soft X-ray observations, but also is capable of successfully reproducing the absorption lines from the highly ionized Fe ions. In the recent X-ray observations of many AGNs it is evident that many absorptions lines with relativistic velocities $\sim$ few percent of light velocity are present in the spectra [7–9].

To check the consistency of our model we should fit observations with our transmitted spectra. For this purpose, we have computed the grid of models and constructed the CP WA tables *cpwabs* in the FITS (Flexible Image Transport System) format. This task was made in the frame of European Union Seventh Framework Program (FP7/2007-2013) under grant agreement No. 312789, and are ready to be fitted with the observed high resolution X-ray data (http://stronggravity.eu/results/models-and-data/). Detailed line fitting for WA is not included in this thesis but the work is in progress. In return, such large grid of CP models allowed us to make systematic studies of AMDs for different spectral shapes of illuminated radiation, and put final conclusions on their shapes in the Section below.

4.2 Universal AMD Shapes

In the previous Sect. 4.1, I showed that the CP assumption of the absorbing clouds in the TITAN computations successfully recovers the observed nature of the AMD in the Sy1.5 galaxy Mrk 509. Nevertheless, the resulting normalization of the model is a factor 30 higher than the observed AMD for this object. In addition to Mrk 509, upto now six other AGNs: NGC 3783, NGC 5548, MCG-6-30-15, NGC 3516, NGC 7469 and IRAS 13349+2438 are known, for which the AMD is determined from high resolution X-ray observations [10] (see Fig. 2.4). The overall normalization of AMDs for these six objects as obtained from the observations is of the order of $\sim 4 \times 10^{21}$ cm^{-2}. Furthermore, all of the six observed AMDs have shown one prominent discontinuity between $\log \xi \sim 0.8$ and 1.7. This is different from the case of Mrk 509 where Detmers et al. [4] obtained two prominent dips around the ionization degree $\log \xi \sim 2$–3 and 3–4, with slightly lower normalization.

To reproduce the observed AMD theoretically, one should consider continuous ionization structure of WA in such a way that the photoionization computations pass through the broad ionization states. This method was applied recently by Stern et al. [3] assuming a radiation pressure confinement (RPC) of the WA material. Using CLOUDY photoionization code, the authors were successful in reproducing the observed normalization and the slope of AMD for those six objects. However, they were not able to quantitatively reproduce the deep minimum in column density for those 6 objects present in the $\log \xi$ between 0.8 and 1.7 (see Fig. 2.4).

In order to have further insight on AMD models and their universal shapes, it is necessary to answer some questions that are releavent in this research. The very first question is how we can explain the discrepancy between the observed and modelled AMD normalizations? Next: what causes the variation in the number of AMD dips for different objects i.e., two prominent dips in Mrk 509 and one prominent dips in other 6 Sy1 galaxies at different ionization degrees. I investigate these issues in details in this Section.

I have computed over several hundreds of AMD models, but to make my research clear, all results presented in this Section are for one warm absorber density $n_{\mathrm{H}} = 10^8$ cm^{-3}, unless stated otherwise. Furthermore, I considered the grid of ionization parameters ξ_0 from 10^3 up to 10^5 erg cm s^{-1}, and total column densities from 10^{19} up to 6.3×10^{23} cm^{-2}. The most interesting models are those with total column densities high enough that the matter structure displays broad ionization levels from the total ionization at the illuminated face of the cloud to the almost neutral matter on the back side. And those models are mostly presented in the Section below, unless it is stated.

4.2.1 General SED Shapes

It is well known that the thermal and ionization structure of the WA cloud depends on the shape of the ionizing SED. The study of stability curve for different shapes of the radiation field revealed that the resulting curves are significantly different, in particular very sensitive to the soft UV/soft X-ray part of the SED [2, 11–13]. Also, it has been shown that thermal and ionization states of WA are sensitive to their density if the ionizing continuum is sufficiently soft, i.e. dominated by the UV/soft X-ray [2, 12]. These conclusions link with my result in the Sect. 4.1, where the AMD model with Mrk 509 SED shape depends on the density.

To make systematic studies of AMD, I consider the variation in SED shape assuming that overall spectrum originates from the AGN center. Next, I study how different parameters influence the nature of AMD models computed with CP assumption in TITAN. The SED shapes are computed by including two major components that contribute to the broad AGN spectra i.e., black body disk spectrum and the X-ray power law. The variation in SED shapes is achieved by changing the parameters: mass accretion rate $\dot{m}$, the black hole spin a, and power law photon index Γ. The mass of the SMBH is taken as $M_{bh} = 10^8$ $M_\odot$ in all cases.

Since the incident radiation has already a large number of parameters, and in addition with input parameters of the warm absorber (ξ_0, n_H, N_{tot}), the amount of computed models is very large. I have found that important conclusions on AMD shape can be visible when two groups of SED are considered. Therefore, for clarity of this Section, I present results for two groups of SED which differ by the strength of black body disk spectrum, and by the relative normalization of disk to X-ray power law emission. The first group, named SED A, represents the case dominated by big blue bump component, and it is calculated for $\dot{m} = 0.1$ and $a = 0.1$. The second group, named SED B, has 100 times weaker black body disk component than SED A, and it is calculated for $\dot{m} = 0.001$ and $a = 0.9$.

For the given set of a and $\dot{m}$, the variation in the X-ray power law slope is done by taking ten values of photon index Γ from 1.4 up to 3.2 with 0.2 grid step. While changing the X-ray photon index, the X-ray luminosity is kept constant for the given values of a and $\dot{m}$. This condition is achieved by changing the join between the UV and X-ray part of the spectrum i.e., by changing the α_{OX} parameter, which displays the relative flux emitted at 2500 Å to the flux emitted in soft X-rays at 2 keV, and it is defined by relation

$$\alpha_{OX} = -0.384 \log \left(\frac{F_2 \, keV}{F_{2500 \text{Å}}} \right). \tag{4.1}$$

For group A SEDs, the black body disk component is always more luminous than the X-ray power law, while for group B SEDs, luminosities of both components are equal. The total luminosity depends mostly on the disk accretion rate. The resulting two groups of parameters and corresponding luminosities are presented in Table 4.1. Overall spectral shapes are drawn in Fig. 4.6 for group A, and in Fig. 4.7 for group B.

Table 4.1 The parameters used for computing the general SEDs. The spin a, mass accretion rate $\dot{m}$, disk luminosity L_{disk}, the X-ray luminosity L_X and the bolometric luminosity L_{tot}, photon index Γ are presented in columns 2, 3, 4, 5 and 6 respectively. The mass of the SMBH in all cases is taken as $M_{\mathrm{bh}} = 10^8\ M_\odot$

SEDs	a	$\dot{m}$	L_{disk}	L_X	L_{tot}	Γ	α_{OX}
A	0.1	0.1	1.257e+45	1.257e+43	1.269e+45	1.4–3.2	−1.9 to −2.8
B	0.9	0.001	1.250e+43	1.257e+43	2.507e+43	1.4–3.2	−1.3 to −2.1

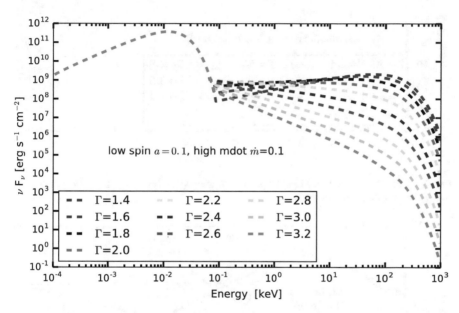

Fig. 4.6 SED A shapes obtained for the parameters $a = 0.1$ and $\dot{m} = 0.1$. Cases with different X-ray photon indexes are shown

Depending on the ratio of luminosities of two SED's components, group A has $L_{\mathrm{disk}}/L_X = 100$ and it is dominated by strong optical UV emission, and the second group B has $L_{\mathrm{disk}}/L_X = 1$, with 100 times weaker optical UV bump, which is a consequence of lower accretion rate. Nevertheless, the overall X-ray continuum has lower normalization in case A in comparison with case B spectra. All above differences between SED of both groups A and B are presented in Fig. 4.8.

4.2.2 AMD Dependence on X-Ray Power Law Slope

In the aim to study how a slope of the X-ray power law component affects the resulting AMD models, I fixed parameters for each cloud: $n_{\mathrm{H}} = 10^8\ \mathrm{cm}^{-3}$, $\xi_0 = 1.36 \times 10^4$ erg cm s^{-1} whereas the N_{H} is varied until the lowest possible temperature is reached.

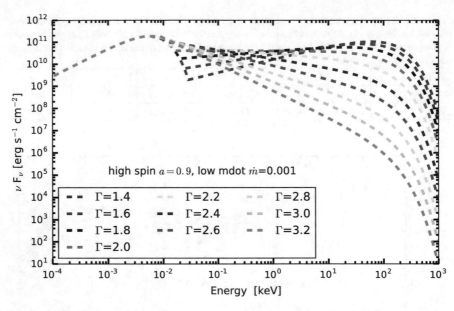

Fig. 4.7 SED B shapes obtained for the parameters $a = 0.9$ and $\dot{m} = 0.001$. Cases with different X-ray photon indexes are shown

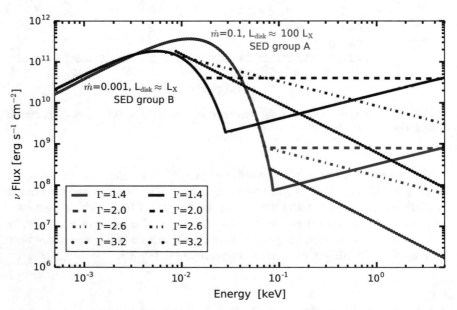

Fig. 4.8 Two types of SEDs: SED A (red lines) and SED B (black lines) are shown for comparison. For clarity only few cases of Γ values are displayed here

Only models with broad ionization distribution are shown for various values of X-ray power law photon index Γ. The temperature variations across the depths of the clouds for group A SEDs are shown in the Fig. 4.9. The Compton temperature differes by ~ 1 order of magnitude between the models which differ by the extreme Γ values i.e. 1.4 and 3.2. With increasing value of Γ, the amount of the sudden temperature drop changes from 2 to 1. For $\Gamma \leq 2.0$, there are two distinct thin regions where a quick drop in the temperature happens. However for $\Gamma > 2.0$, only a single temperature drop is seen.

The features of sudden drop in the temperature are also nicely reflected by the switch of the number of the prominent AMD dips from 2 to 1 in the corresponding AMD models for the selective Γ values as shown in the Fig. 4.10. The models clearly demonstrate that the number of AMD dips depends on the slope of the X-ray power law used in the illumination. This effect is directly reflected in the value of temperature at the cloud surface which is derived from energy balance equation. According to expectations, more energetic photons do heat gas stronger in photoionization and

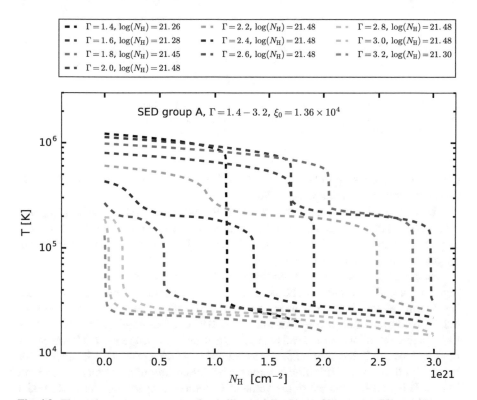

▪ ▪ $\Gamma = 1.4$, $\log(N_{\mathrm{H}}) = 21.26$	▪ ▪ $\Gamma = 2.2$, $\log(N_{\mathrm{H}}) = 21.48$	▪ ▪ $\Gamma = 2.8$, $\log(N_{\mathrm{H}}) = 21.48$
▪ ▪ $\Gamma = 1.6$, $\log(N_{\mathrm{H}}) = 21.28$	▪ ▪ $\Gamma = 2.4$, $\log(N_{\mathrm{H}}) = 21.48$	▪ ▪ $\Gamma = 3.0$, $\log(N_{\mathrm{H}}) = 21.48$
▪ ▪ $\Gamma = 1.8$, $\log(N_{\mathrm{H}}) = 21.45$	▪ ▪ $\Gamma = 2.6$, $\log(N_{\mathrm{H}}) = 21.48$	▪ ▪ $\Gamma = 3.2$, $\log(N_{\mathrm{H}}) = 21.30$
▪ ▪ $\Gamma = 2.0$, $\log(N_{\mathrm{H}}) = 21.48$		

SED group A, $\Gamma = 1.4 - 3.2$, $\xi_0 = 1.36 \times 10^4$

Fig. 4.9 Thermal structure across the clouds illuminated with the SED A, for different values of Γ in the range 1.4–3.2. The structures are computed for the various values of N_{H} until the lowest temperature is reached. The values of $n_{\mathrm{H}} = 10^8$ and $\xi_0 = 1.36 \times 10^4$ are fixed in these computations

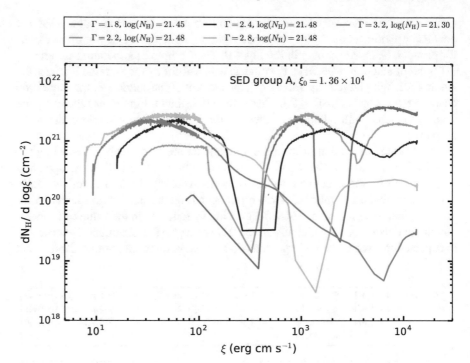

Fig. 4.10 AMD models for the SED group A for the various Γ values. The AMDs are computed for all the models with various values of N_H until the broad ionization distribution is obtained. The values of $n_H = 10^8$ cm^{-3} and $\xi_0 = 1.36 \times 10^4$ erg cm s^{-1} are used in the model computations

Compton or free-free processes. But, the position of those drops in case of SED A does not agree with the observed AMDs.

The common feature seen in the nature of all AMDs for varying Γ values of group A models is that the overall normalization always remains at the level of $\sim 4 \times 10^{21}$. This result is in very good agreement with the AMD normalization obtained for six Sy galaxies from the observations [10].

The temperature structure in the case of SED B models for the increasing photon index in range $\Gamma = 1.4$–3.2 is shown in the Fig. 4.11. From the above figure, a clear separation between two types of models is seen. For the Γ values in the range 1.4–2.2, high amount of absorbing material $\geq 2 \times 10^{23}$ cm^{-2} is required in order to pass through the broad ionization levels to reach the minimum temperature. However, for the Γ values >2.2, the column density required is $\leq 10^{23}$ cm^{-2}. This is obvious for the reason that, SEDs with high Γ values have less X-ray photons than SEDs with low Γ. This fact is reflected in the value of Compton temperatures for each model where the temperature decreases with increasing Γ.

The AMDs corresponding to SED B clouds, for the selected values of Γ, are given in Fig. 4.12. With increasing photon index of the X-ray power law used in the illumination, the prominent AMD dips reduce from two ($\Gamma = 1.8$) to one ($\Gamma = 2.6$).

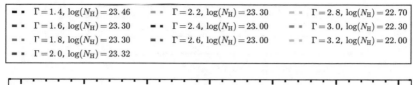

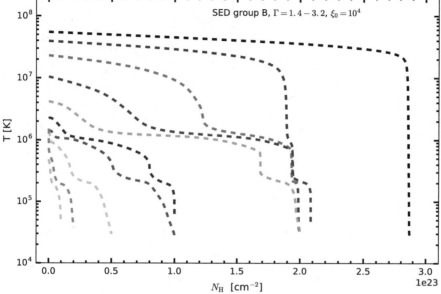

Fig. 4.11 Thermal structure across the clouds developed when it is illuminated with the SED B, $\Gamma = 1.4$. The structures are computed for the various total column densities of the clouds all having the ionization parameter at the incident surface equal to $\xi_0 = 1.7 \times 10^4$ erg cm s^{-1} and the gas density $n_H = 10^8$ cm^{-3}

Furthermore, the position of low ionization dip slightly moves toward lower value of ionization parameter, and it may explain observations, but the overall normalization is too high in contradiction with observations.

Note that, in case of both groups of SEDs A and B, there are models for $\Gamma = 1.4$ which display strong temperature drop across the whole cloud. The resulting AMD cannot be constructed since there is not enough absorption material for each ionization parameter. This is according to my expectations that for hard X-ray power law the thermal instability is the strongest and such cloud cannot survive in any stationary equilibrium.

The most important conclusion of this Section is that for models dominated by strong optical UV bump and weak X-ray component, i.e. SED A, the AMD normalization derived with CP models agrees with observations of six Seyfert galaxies. When the disk emission is equal or smaller than X-ray power law of illuminated continuum, i.e. SED B, the AMD normalization is one order of magnitude higher than the one predicted by observations.

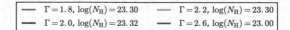

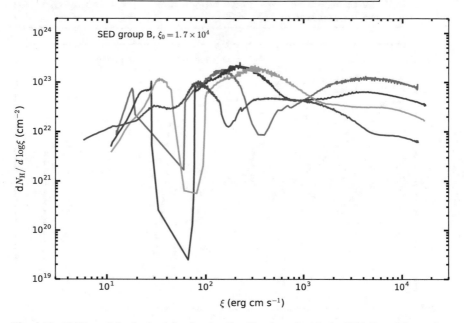

Fig. 4.12 AMD models obtained for the absorber illuminated with the SED B, and for various Γ values. The models are computed for the ionization parameter at the incident surface equal to $\xi_0 = 1.7 \times 10^4$ erg cm s^{-1} and the gas density $n_H = 10^8$ cm^{-3}

4.2.3 AMD Dependence on Initial Ionization Parameter

In case of high ionization parameter the illuminated cloud surface is hot, of the order of Compton temperature. In such a case, when ionization parameter decreases with depth, the temperature structure displays two steep drops as it is presented in Fig. 4.13 for case A SED type, $\Gamma = 2.0$ and $\xi_0 = 1.36 \times 10^4$ erg cm s^{-1}. For lower ionization parameter at the cloud surface, the outer layers are not so hot and overall temperature structure displays one steep drop, as it is shown in Fig. 4.14, for the same SED, but $\xi_0 = 1.36 \times 10^3$ erg cm s^{-1}.

For group A spectrum, the models with lower ionization parameter contain only one thermally unstable region since the temperature at the surface of the cloud is below the first unstable region present in the cases for higher ionization parameter. This feature is nicely reflected in the AMD natures shown in the Figs. 4.15 and 4.16. But the observed one dip in AMDs of six objects is located for log ξ between 0.8 and 1.7 which is in disagreement with my model. Theoretical AMD, shown in Fig. 4.16, possesses one dip for log ξ between 2 and 2.5.

To explore the dip's origin, the comparison of the major gas heating-cooling processes i.e., Compton and the free-free processes in different models considered is

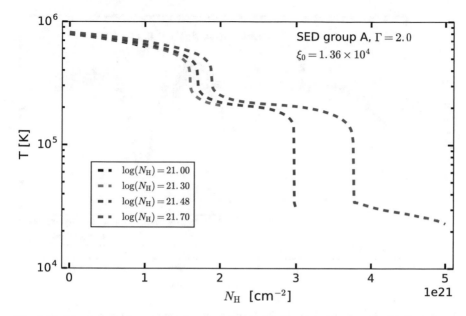

Fig. 4.13 Thermal structure across the cloud developed when it is illuminated with the SED A, $\Gamma = 2.0$. The structures are computed for $n_H = 10^8$ cm^{-3} and various values of N_H. The ionization parameter at the surface of the cloud is equal to 1.36×10^4 erg cm s^{-1}

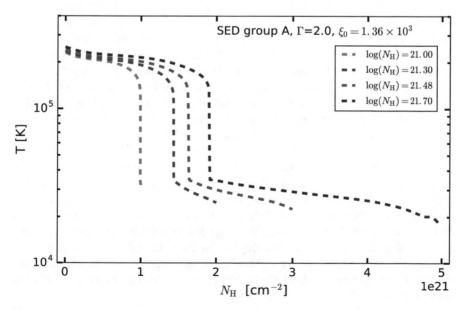

Fig. 4.14 Thermal structure across the cloud developed when it is illuminated with the SED A, $\Gamma = 2.0$. The structures are computed for $n_H = 10^8$ cm^{-3} and various values of N_H. The ionization parameter at the surface of the cloud is equal to 1.36×10^3 erg cm s^{-1}

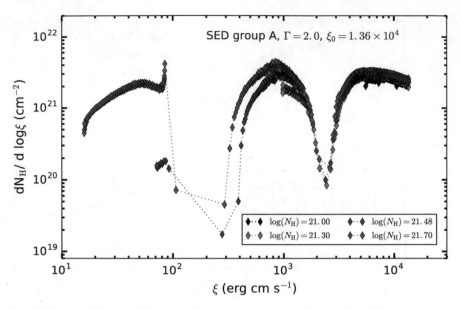

Fig. 4.15 AMD models for SED A, for the same model parameters as in Fig. 4.13

shown in the Fig. 4.17. It is clear from the figure that when the ionization degree is higher i.e., $\xi_0 = 1.36 \times 10^4$ erg cm s^{-1} (top panel), the Compton contribution to the cooling (red dashed line) is comparable to the free-free cooling (blue dashed line) at the surface of the gas cloud. The situation differs as we move to the lower ionization parameter $\xi_0 = 1.36 \times 10^3$ erg cm s^{-1} (bottom panel) where the Compton cooling contribution decreases by an order of magnitude. At the ionization parameter corresponding to the positions of the AMD dips, the sudden changes in the heating-cooling rates are also observed as demonstrated in the Fig. 4.17. These positions correspond to the sudden temperature drops shown in the thermal structures in the Figs 4.13 and 4.14 respectively. In case of the lowest ionization parameter considered in this case (bottom panel of the Fig. 4.17), there is only one sudden change present. These results are again the confirmation that the AMD minima are caused by thermally unstable regions where the abrupt change in the heating-cooling processes also happens.

Since AMDs from group B SEDs are always too high in normalization, I expect that thay will never fit the observed AMD for six Sy galaxies. Futhermore, I have checked, that those models do not depend on the value of ionization parameter significantly. My general conclusion from this Section is that the ionization parameter defined at the illuminated cloud surface changes the amount of dips in AMD, but does not influence the position of those dips.

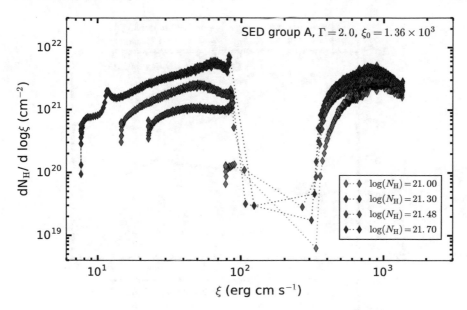

Fig. 4.16 AMD models for SED A, for the same model parameters as in Fig. 4.14

4.2.4 AMD Density Dependence

Here, I investigate how the gas density defined at the irradiated face of the cloud influences the AMD structure. For this purpose, all considered models are computed with the SED A, $\Gamma = 2.0$, and the same ionization parameter $\xi_0 = 1.38 \times 10^3$ erg cm s^{-1}. The only parameter which has been changed is the value of density at the illuminated cloud surface. I made calculations for various values of $n_H = 10^8$, 10^{10}, 10^{11} and 10^{12} cm^{-3} respectively.

The cloud temperature structures and the corresponding AMD models are presented in Figs. 4.18 and 4.19 respectively. There is no significant difference in the overall nature of the AMD for $n_H \leq 10^{11}$ cm^{-3}. For these models, only one AMD dip is present since the initial value of ionization parameter is already below the value at which the first dip occurs. However, the depth of the AMD dips as well as the normalizations are slightly different even for these range of densities. With increase in the gas density, the depth of the AMD dip decreases and the overall normalization increases as seen from the Fig. 4.19. As the gas density becomes 10^{12} cm^{-3}, the AMD dip is completely diminished and a continuous AMD is obtained for such model. From the temperature structure one may expect that the strong temperature drop which causes strong decrease of AMD at $\xi \sim 10$ is caused by thermal instability, but more high density models should be calculated to confirm this hypothesis.

For the purpose of this thesis, I can check high density physical behaviour by plotting the contribution of Compton and free-free processes in the heating and cooling of the gas. The importance of Compton heating in comparison to free-free

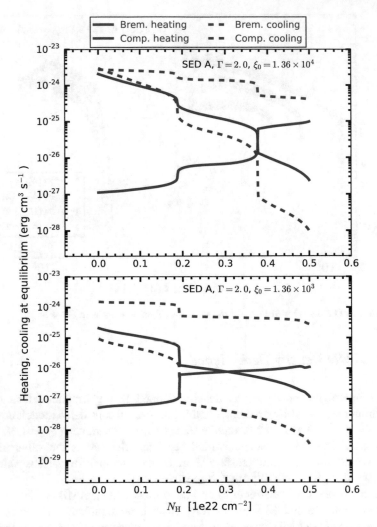

Fig. 4.17 The free-free and Compton contribution to the heating-cooling of the irradiated gas. The effect of the change in ionization parameters in our model is shown. All models are computed with $n_H = 10^8$ cm^{-3} and $N_H = 5 \times 10^{21}$ cm^{-2}

heating for each model with different gas densities is shown in Fig. 4.20. From the above figure, the physical reason for this AMD behaviour is visible since for gas at lower density 10^8, there is a clear domination of Compton heating over free-free heating. When density of the cloud surface increases, the dominance of free-free heating over the Compton heating is clearly visible inside the cloud. When $n_H \geq 10^{11}$ cm^{-3}, this domination occurs across the whole absorber. A similar result was obtained for the WA model with gas density $n_H = 10^9$ cm^{-3} when the Mrk 509

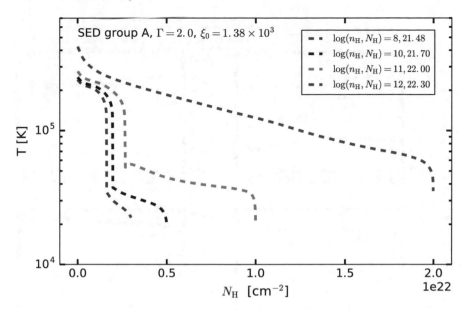

Fig. 4.18 Thermal structure across the cloud defined by the parameters as shown in the figure. The four cases of n_H at the illuminated face of the cloud is considered. These computations are done for the SED case A and $\xi_0 = 1.38 \times 10^3$ erg cm s^{-1}

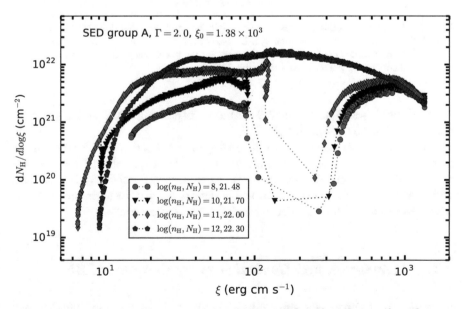

Fig. 4.19 AMD models computed for the gas densities $n_H = 10^8$, 10^{10}, 10^{11} and 10^{12} cm^{-3} respectively. The temperature structure for the corresponding models are shown in the Fig. 4.18

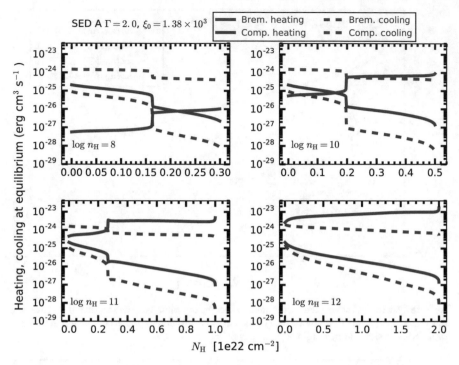

Fig. 4.20 Heating and cooling processes operating in the gas clouds described by the parameters given in the Fig. 4.18. When the gas density is increased to 10^{12} cm^{-3}, the free-free heating-cooling processes become more significant than in the cases for the low density values. As a result of this behaviour there is no any significant drop in the AMD

SED was used (see Figs. 4.1 and 3.9 for the corresponding AMD and the heating-cooling mechanisms respectively).

There is a clear evidence that for increasing density the dominant heating mechanism changes from Compton heating to free-free heating. This fact may be connected with wind launching mechanisms. When density is relatively low, we deal with Compton heated winds, but for higher density we may meet thermally driven winds. Nevertheless, this hypothesis needs further considerations.

4.2.5   Comparison with General Shape of Observed AMD

In Sect. 4.1.1, I compared the observed AMD and our CP absorber model for Mrk 509 where an excellent agreement on the positions of the AMD dip is found. Nevertheless, the best fit AMD normalization was higher by the factor ~30 than the observed one found by Detmers et al. [4]. Furthermore, I found that modelled AMDs determined for group A SED have normalization in agreement with observations. AMDs from

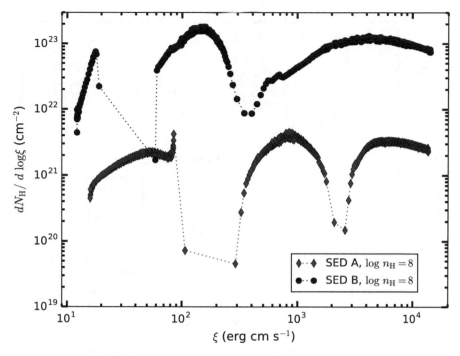

Fig. 4.21 A comparison between the AMD models computed for two spectral shapes: SED A (red) and SED B (black) for the parameters: $n_H = 10^8$ cm^{-3} and $\xi_0 = 1.35 \times 10^4$ erg cm s^{-1}. The figure is reproduced from Adhikari et al. [14] with the reprint permission from the Polish Astronomical Society

clouds illuminated by B type SEDs have a few tens higher normalization, which is clearly shown in Fig. 4.21.

To understand what physical parameters shape the observed AMD, I have investigated photoionization computations of high density clouds illuminated by A type SEDs. In Fig. 4.22, I show a comparison between the observed AMDs for 6 Seyfert galaxies published by Behar [10] and our best fit model computed with the general SED shape A, and gas density 10^{12} cm^{-3}. The comparison shows that there is a very good agreement in the overall AMD normalizations ($\sim 4 \times 10^{21}$) of those sources and our model. Nevertheless, for this high density our computations stop at the ionization parameter $\xi \sim 4.5$ erg cm s^{-1}. Therefore, only a part of the AMD drop is reconstructed. The models with the further increase of N_H do not converge in TITAN as the cloud is already optically very thick and reaches to the convergence limit of the TITAN code.

Previously, the AMD modelling was done by Stern et al. [3] assuming that the WA is compressed by the radiation pressure using the numerical code CLOUDY. However their RPC models did not reproduce the AMD dips which are caused due to the thermally unstable regions operating in the AGN outflow. In our opinion, this is related to the issue that the transfer of radiation in CLOUDY is done using an escape

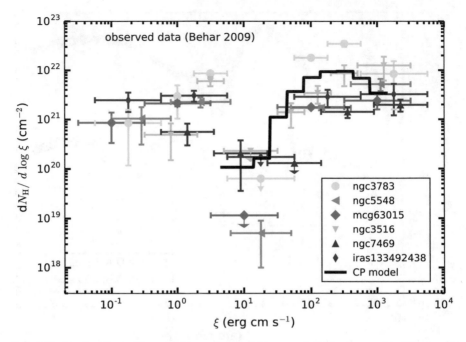

Fig. 4.22 The best fit AMD model (shown by black histogram) obtained using the SED A with CP assumptions in TITAN overplotted with the observed AMD points obtained by Behar [10] for six Sy1 galaxies. The AMD model is obtained for the parameters: SED A ($\Gamma = 2.0$) and the gas density $n_H = 10^{12}$ cm^{-3}. This clearly demonstrates that the Seyfert outflow can be well explained by the WA under CP

probability method. While computing the CP clouds structure, this method does not calculate the local radiation pressure. However in TITAN, the radiative transfer is done using the ALI method where the proper source function term in the radiative transfer equation is taken into account as described in the Chap. 2.

In ALI method, the radiation pressure and then the accurate ionization parameter structure can be precisely calculated. Nevertheless, the method is very rigorous in terms of the treatment of any structure gradients. If those gradients are too steep the code does not converge due to the fact that the determinants of the matrix in the radiative transfer solution become infinity. Therefore, with the TITAN code we can only find the instability, but we cannot study this region.

Summarizing, the agreement in the overall shape of the modelled and observed AMD is very encouraging and clearly demonstrates that the CP models are very successful in explaining the WA properties in Sy galaxies.

Our result for high density warm absorber puts tighter constraints on the wind location. Assuming typical Sy1 luminosity to be 10^{45} erg s^{-1}, the location of such dense cloud could be estimated at $r \sim 10^{15}$ cm which is about 30 gravitational radii for 10^8 M$_\odot$ black hole mass. This is very close to the SMBH and the question arises if such a warm absorber can survive in very changeable innermost AGN region?

X-ray magnetic flares above an accretion disk were proposed by several groups [15, 16], which may be a mechanism for hot ionized gas cloud formation. Nevertheless, the estimation of lifetime for such clouds is rather not optimistic. From the work by [17], flares can live only 10 orbital periods, which gives the lifetime of the order of one year. In such situation the warm absorber will be highly variable and unstable, in contradiction to what we really observe in 50% of AGNs.

4.3 Conclusions

In this Chapter, I studied the WA clouds present in the AGN outflow by performing the photoionization simulations using the numerical code TITAN developed by [18]. The various types of SEDs are considered to illuminate the warm absorber with an assumption that the total pressure inside the cloud i.e., $P_{gas} + P_{rad}$ is constant. The temperature and ionization structures are computed to obtain the AMD for each models. From the analysis of AMDs derived for various combinations of parameters, I summarize the main conclusions below.

1. I computed AMD for the case of Mrk 509, for which SED is well known from the multi-wavelength long term monitoring. I found discontinuities in AMD in the range of the ionization parameter, $\log \xi = 2–3$ and $\log \xi = 3–4$, similar to that obtained observationally by Detmers et al. [4]. Such observed discontinuity was often interpreted as the absence of ions in the thermally unstable regions [19], which are also visible in the stability curve presented in the Fig. 3.8. This observed minimum is also described as two geometrically distinct regions along the line of sight representing high ionization region and low ionization region [19]. Contrary to the result obtained by Detmers et al. [4], instead of two discrete absorbing zones, this work explains AMD as a continuous single absorber under CP.
2. In case of Mrk 509 SED, I found that the AMD depends on the gas density n_H. For a high value of gas density i.e., when $n_H = 10^9$ cm^{-3}, only one AMD dip becomes significant. The best fit model with gas density 10^8 cm^{-3} successfully reproduces the observed positions of the AMD dips. All the WA models for Mrk 509 SED require higher values of the column density in order to produce the broad ionization range which has been deduced from the X-ray observations. As a result of this, the normalization of the modelled AMD is higher by the factor $\sim$30 than the observed one, for our best fit model.
3. For the best AMD model, the location of the warm absorber in case of Mrk 509 obtained from my analysis is $r = 1.39 \times 10^{16}$ cm, about 10–100 times closer to the nucleus than the upper limit for the wind location found by Kaastra et al. [5] for 5 ionization components from variability method. Despite this discrepancy, I present here the first estimation of wind location done with the use of AMD for Sy1.5 bright AGN. My result agrees with the position of BLR derived by reverberation mapping technique which is of the order of 10^{16} cm. Therefore, it

confirms the scenario that the warm absorbers are connected with BLR regions as suggested by the first results of the high resolution data [20, 21].

4. I studied the influence of the SED shapes on the AMD nature by taking significantly different radiation types SED A and SED B. SED A and SED B differ by the ratio of optical/UV flux to the X-ray flux. I found that when the absorbing cloud under constant pressure is illuminated by the SED A, the small amount of the material with $N_H < 10^{22}$ cm^{-2} is sufficient to produce the broad ionization structure covering ~ 4 orders of magnitude. This produces the average AMD normalization in the range 10^{21} to 10^{22} which is in excellent agreement with the observationally obtained average AMD normalization for 6 Sy1 galaxies by Behar [10].

5. Moreover, due to the fact that SED A is dominated by optical/UV component, I found that AMD models depend on the gas density n_H used in the computations. For the gas density 10^{12} cm^{-3}, I found the AMD model that is best fit to the AMD shape obtained for the 6 Sy1 galaxies by Behar [10].

6. High density warm absorber puts constraints on the wind location. Assuming typical Sy1 luminosity to be 10^{45} erg s^{-1}, the location of such dense cloud could be estimated at $r \sim 10^{15}$ cm which is about 30 gravitational radii for a black hole of 10^8 M$_\odot$. Up to now, there is a theory explaining the origin of such a dense wind. Furthermore, the stability of such dense cloud should be explored in the future studies.

The models presented in this Chapter demonstrate that the CP assumption in the WA modelling successfully produces the observed AMD in various AGNs. Moreover, the dependence of the AMD dips on the gas density may give an independent way of diagnosing the density of the ionized outflow in AGNs for which the observed AMDs exist.

References

1. Adhikari TP, Różańska A, Sobolewska M, Czerny B (2015) ApJ 815:83
2. Różańska A, Kowalska I, Gonçalves AC (2008) A&A 487:895
3. Stern J, Behar E, Laor A, Baskin A, Holczer T (2014) MNRAS 445:3011
4. Detmers RG et al (2011) A&A 534:A38
5. Kaastra JS et al (2012) A&A 539:A117
6. Adhikari TP, Różańska A, Sobolewska M, Czerny B (2016) In: Różańska A, Bejger M (eds) 37th meeting of the polish astronomical society, vol 3, pp 239–242
7. Markowitz A, Reeves JN, Braito V (2006) ApJ 646:783
8. Dauser T et al (2012) MNRAS 422:1914
9. Tombesi F, Cappi M, Reeves JN, Nemmen RS, Braito V, Gaspari M, Reynolds CS (2013) MNRAS 430:1102
10. Behar E (2009) ApJ 703:1346
11. Różańska A, Goosmann R, Dumont A-M, Czerny B (2006) A&A 452:1
12. Chakravorty S, Kembhavi AK, Elvis M, Ferland G (2009) MNRAS 393:83
13. Chakravorty S, Misra R, Elvis M, Kembhavi AK, Ferland G (2012) MNRAS 422:637
14. Adhikari TP, Różańska A, Hryniewicz K, Czerny B (2018) In: Różańska A (ed) XXXVIII polish astronomical society meeting, vol 7, pp 322–325

15. Czerny B, Goosmann R (2004) A&A 428:353
16. Ponti G, Cappi M, Dadina M, Malaguti G (2004) A&A 417:451
17. Goosmann RW, Czerny B, Mouchet M, Ponti G, Dovčiak M, Karas V, Różańska A, Dumont A-M (2006) A&A 454:741
18. Dumont A-M, Abrassart A, Collin S (2000) A&A 357:823
19. Holczer T, Behar E, Kaspi S (2007) ApJ 663:799
20. Kaastra JS, Steenbrugge KC, Raassen AJJ, van der Meer RLJ, Brinkman AC, Liedahl DA, Behar E, de Rosa A (2002) A&A 386:427
21. Gabel JR et al (2003) ApJ 583:178

Chapter 5
Intermediate Line Region in AGNs

Abstract In this chapter, I present that the gas density is also of great importance when searching emission lines in AGNs. With the use of photoionization modelling, I extend the classical (Netzer and Laor ApJ 404:L51, 1993 [1]) model for BLR and NLR in AGN, and I show the existence of ILR which is in agreement with the recent observations. Furthermore, I study how the ILR properties depend on different types of AGN and on different cloud parameters.

All the models of the emission line clouds are computed using the numerical code CLOUDY which is described in the Chap. 2 of this thesis. Note that, Netzer and Laor [1] have used different photoionization code ION [2], which is similar to the CLOUDY's numerical scheme. In Sect. 5.1.5, I show that the canonical model of Netzer and Laor [1], where the observed apparent gap in the line emission region between BLR and NLR is explained by the dust content in NLR clouds, is fully reproduced with CLOUDY code for the same physical parameters of radially distributed clouds. Furthermore, by considering five different spectral shapes, in addition to the one general Sy1 SED [3] assumed by Netzer and Laor [1], I show that the above result does not depend on spectral shape.

In the next section I show that the observed properties of ILR can be explained by increasing gas density at the sublimation radius. Such result does not depend on the adopted SED and overall radial density profile. In addition for such high density, CP models give the same line luminosities as CD models. Final results from the CLOUDY modelling obtained for different AGN types are compared and discussed with the observationally derived properties of the emission line regions in those sources.

© Springer Nature Switzerland AG 2019
T. P. Adhikari, *Photoionization Modelling as a Density Diagnostic
of Line Emitting/Absorbing Regions in Active Galactic Nuclei*,
Springer Theses, https://doi.org/10.1007/978-3-030-22737-1_5

5.1   Model Set Up and the Parameters

To portray the general picture of the line emitting regions, I consider a continuous distribution of optically thick spherical gas clouds above an AD, placed at different radial distances extending from the BLR out to the NLR. Each cloud at a given radial distance r from the nucleus represents the gas in the emission region described by the parameters: hydrogen number density, n_H, dimensionless ionization parameter U, total hydrogen column density, N_H, and the chemical abundances (each of them are defined in the Chap. 2). The distance to the emission cloud from the central engine of the AGN, the gas density and hydrogen ionizing photon flux, related by an Eq. 2.12, determines the ionization degree of the gas. In general the considered clouds are in CD assumption except in the cases where I explicitly mention the use of CP clouds. In the section below, I present the description of the parameters adopted in the computations of the line emission models.

5.1.1   Gas Density Profiles

For the investigation of the global emission properties of the gas present at large distance scale, the approximation that the cloud density is a function of radial distance given by power law is adequate as shown by Netzer and Laor [1]. In the first part of this chapter, I adopted this distribution of the density parameter

$$n_H = A \, (r/R_d)^{-\beta}, \tag{5.1}$$

where β is the slope that defines the steepness of the power law. A is the density normalization factor. The dust sublimaton radius $R_d = 0.1 \, \text{pc}, \beta = 1.5$ and $A = 10^{9.4}$ cm^{-3} in Eq. 5.1 correspond to the Netzer and Laor [1] canonical density profile. I investigate the line emission properties by varying the slope β and the normalization A in our models.

The above assumption about gas density is relaxed in Sect. 5.4, where I consider more realistic density profiles expected at the upper part of the AD atmosphere. Such statement is consistent with the concept of AGN disk winds, when the emitting clouds are formed above the disk by mechanism which is still not fully solved. But magnetic, radiation and thermal pressure are under consideration to power disk winds. To determine the realistic cloud radial density profile, the AD vertical structure is computed assuming the standard thin accretion disk. The total luminosity, Eddington ratio and the mass of the black hole are parameters in those computations (see Różańska [4], for full model discription). The radiative transfer is solved with the diffusion approximation, and the density of our interest is the density of upper atmosphere at the optical depth $\tau = 2/3$. Such value of density computed at each radius is taken to be the initial cloud density, which is reasonable when the atmospheric gas forms the disk wind. The exact values of cloud density profiles, adopted in this scenario, are given in Sect. 5.4.

5.1.2 Total Column Density of the Clouds

In all the CLOUDY models for the emission gas presented in this chapter, I adopt the column density profile as a function of radial distance in the form

$$N_{\rm H}(r) = 10^{23.4} \, (r/R_{\rm d})^{-1}. \tag{5.2}$$

I took the normalized value of the column density of the cloud located at the sublimation radius $R_{\rm d}$, as $N_{\rm H} = 10^{23.4} \, {\rm cm}^{-2}$, as in canonical [1] model.

5.1.3 Chemical Abundances and the Dust Sublimation Radius

Since the separation between BLR and NLR in AGN was explained by the presence of dust [1] at the certain radius, I consider two types of the chemical abundances. For the gas clouds with location smaller than $R_{\rm d}$ the *Solar abundances* are used. Whereas for the clouds with radii $\geq R_{\rm d}$, the *ISM abundances* are used (see Table 2.1 for details). The assumed Solar composition for $r < R_{\rm d}$ has two orders of magnitude higher iron abundance than ISM composition for $r \geq R_{\rm d}$. All other elements display the same magnitude abundances when changing from Solar to ISM composition. The change of abundances mimics the depletion of metals due to dust sublimation as assumed by Netzer and Laor [1]. This ensures the presence of the dust for the distances larger than the dust sublimation radius.

In this chapter, I employ the value of dust sublimation radius $R_{\rm d}$ in two ways. When I derive the line luminosity using the power law prescription of the density profiles (see Eq. 5.1), the value of $R_{\rm d}$ is taken as 0.1 pc. However, I compute $R_{\rm d}$ self consistently using the Eq. 2.17 given by Nenkova et al. [5] for the cases when the realistic density prescription expected in the upper part of the standard AD atmosphere is used. The expression in Eq. 2.17 simply indicates the radius at which, for a given luminosity, the gas temperature reaches the value of 1400 K. Below this temperature, dust can survive as a substantial gas component.

5.1.4 SED Shapes, Luminosities and Ionization Parameter

In order to account for the various types of AGN, I use different spectral shapes of the radiation which enters the radially distributed clouds. All SEDs used in my photoionization simulations are presented in Fig. 5.1. These SEDs are chosen primarily for two reasons: (i) they are recently constrained by combining the data from multi-wavelength observations (except the Band SED) and (ii) this family of SED represents possible types of SED shapes that an AGN can have in general. These

SEDs are incorporated in order to account for the various types of AGN and the differences in the various components of the broad SED.

To represent the general shape of AGN SED, I also considered the Band function $f(E)$ [6] which combines two power laws smoothly by the following relation:

$$
\begin{aligned}
f(E) &= \text{norm.} \left[\frac{E}{100} \right]^{\alpha} e^{\left[\frac{-2(E+\alpha)}{E_{\text{p}}} \right]} \quad \text{for } E < \frac{(\alpha - \gamma) E_{\text{p}}}{(2 + \gamma)}, \\
&= \text{norm.} \left[\frac{(\alpha - \gamma) E_{\text{p}}}{100(2 + \alpha)} \right]^{(\alpha - \gamma)} \left(\frac{E}{100} \right)^{\gamma} e^{(\gamma - \alpha)} \\
&\qquad \text{for } E \geq \frac{(\alpha - \gamma) E_{\text{p}}}{(2 + \gamma)},
\end{aligned} \tag{5.3}
$$

where E_{p} is the peak energy at which the two power laws with slopes α and γ combine smoothly. I choose the slope of the first power law $\alpha = 0.51$ with exponential cutoff at 13.3 eV and combine it with the second power law with slope $\gamma = -1.5$ to produce the Band shape shown by blue line in the Fig. 5.1.

I considered the Sy1.5 galaxy Mrk 509 for which a broad band SED (red dashed line in the Fig. 5.1) derived from the multi-wavelength observation campaigns is available [7], the details of which are given in the Sect. 3.1.

NGC 5548 is among the very well studied AGN. In this study, I used the SED shape of NGC 5548 (green solid line in the figure) derived by Mehdipour et al. [8]. This SED model was constructed by fitting the observational data points obtained using various instruments: stacked *XMM-Newton* data from 12 observations and corresponding simultaneous NuSTAR, INTEGRAL, HST COS observations collected in 2013. Note from the Fig. 5.1 that the soft X-ray below 1 keV dominates in Mrk 509 whereas in NGC 5548 the SED is dominated by harder photons above 2 keV.

The SED of NLSy1 galaxy PMN J0948+0022 is taken from D'Ammando et al. [9]. The data were collected using the Fermi-LAT spectrum, optical, UV, and X-ray data collected by *Swift* (XRT and UVOT). To account for the thermal emission and soft X-ray excess, the disk emission and the Compton scattering by an optically thin thermal plasma near the AD was also included. Figure 5.1 shows that the SED of this source lacks the emission around $\sim$0.1 keV and has an excess of harder photons.

To account for the broad AGN types, the SED of LINER NGC 1097 (black solid line in the figure) is taken from Nemmen et al. [10]. The observational properties of the low-luminosity AGNs (LLAGNs) are significantly different from those of high luminosity AGNs. The LLAGNs do not have the prominent thermal continuum emission in the UV ('big blue bump') which indicates that these sources do not have optically thick geometrically thin accretion disks [11–13].

In order to compare spectral shape of different type of AGN, I normalized all SEDs to $L_{\text{bol}} = 10^{45}$ erg s^{-1} in Fig. 5.1, and this value is adopted in all calculations when power law density radial profile is considered. For more realistic density profiles,

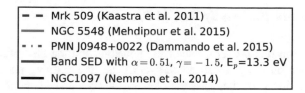

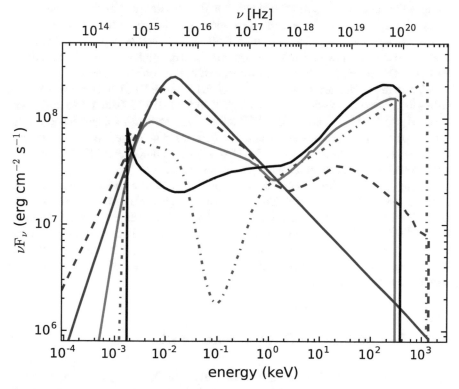

Fig. 5.1 Shapes of the broad band spectra used in my photoionization calculations. All SEDs are normalised to the $L = 10^{45}$ erg s^{-1} and represent the radiation illuminating the cloud at $r = 0.1$ pc. The red dashed line shows Mrk 509, while the solid green line describes NGC 5548. The SED of the NLSy1 PMN J0948+0022 is presented as the magenta dashed dotted line, and the SED produced with the Band function is shown by green solid line. The LINER NGC 1097 SED is shown by the black solid line

luminosities are taken directly from integrations of observations (see Sect. 5.4 for details). The ionized clouds are distributed from $r = 10^{-2}$ up to $r = 10^3$ pc. For the given cloud location and for the given density at the cloud surface, the ionization parameter U is computed by the CLOUDY code using Eq. 2.12.

5.1.5 Line Emissivity Radial Profiles

The emission line properties such as intensities and the line ratios provide crucial
information on the physical conditions of the line emitting plasma in the AGNs.
Using the line ratio analysis, the electron density and temperature, the chemical
composition and the degree of ionization and excitation can all be derived from
observations. A notable feature of AGN spectra is the presence of both high and
low excitation lines. The low ionization lines (LILs) such as Mg II, Fe II and Hβ
are produced in the low ionization regions whereas the high ionization lines (HILs):
[O III], C IV and He II indicate the presence of highly ionized material.

With all the parameters defined in the Sect. 5.1, I derived the line luminosity versus
radius for the major emission lines: Hβ λ4861.36 Å, He II λ1640.00 Å, Mg II λ2798.0
Å, C III] λ1909.00 Å and [O III] λ5006.84 Å and studied its dependence on the
various parameters. I show the existence of ILR based on those lines. Nevertheless,
I include also two additional lines: Fe IIλ (4434.00–4684.00) Å, C IV λ1549.00 Å,
when searching ILR model for wide parameter space.

During the simulations, I consider two conditions: constant density (CP) and the
constant pressure (CP) under which the emission gas operates. The properties of the
single gas cloud under the assumptions of CP and CD and the resulting differences
in the cloud structure are discussed in the Sect. 3.3. When I use CD computations,
the density of a gas cloud at a given radius is kept to a constant value. Whereas in
the CP case, the density values are defined only at the surface of the gas clouds and
it changes across the depths.

5.2 The Condition of ILR Existence in AGN

In this section, I show that CLOUDY code fully reproduces the result of apparent
gap in line emission which naturally separates BLR and NLR in AGN. This effect is
caused by dust content for clouds located further than sublimation radius. In addition,
I check how this result depends on SED shape and the gas density normalization.
To investigate the effect of the various shapes of the incident spectral radiation, I
adopted the Netzer and Laor [1] density prescription to describe the gas clouds in
the emission region. Following their formalism, the gas density normalization A is
set to the value $10^{9.4}$ cm^{-3} at the sublimation radius $R_d = 0.1$ pc in the Eq. 5.1. The
variation in the gas density with the radial location is obtained by setting the slope
$\beta = 1.5$.

Figures 5.2 and 5.3 show the line emission luminosity versus radial distance
derived for the lines Hβ, He II, Mg II, C III] and [O III]. The upper panels of the
Fig. 5.2 show the line luminosities for Mrk 509 (left) and NGC 5548 (right), whereas
the lower panels contain the results for PMN J0948+0022 (left) and Band function
(right). The line emissivity profiles for LINER NGC 1097 is shown in the Fig. 5.3.

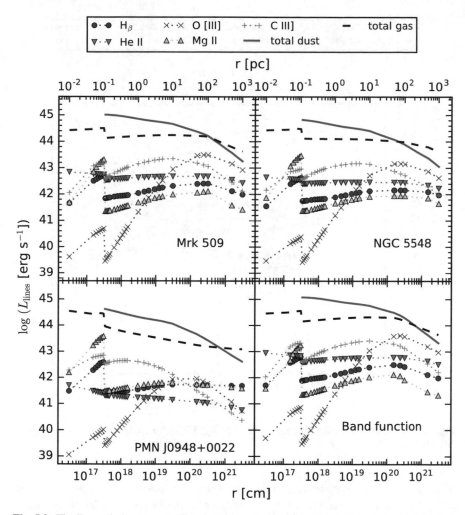

Fig. 5.2 The line emission versus radius for various lines: Hβ (red circles), He II (green triangles down), [O III] (blue crosses), Mg II (cyan triangles up) and C III] (magenta pluses) shown as a function of radius from the illuminating source. For all panels $n_{\rm H} = 10^{9.4}$ cm^{-3} at $r = 0.1$ pc. Total dust emission (magenta continuous line) and the total gas emission (black dashed line) are shown for clarity. Upper left panel presents the results for spectral shapes of Mrk 509, upper right panel- NGC 5548, lower left—NLSy1 PMN J0948+0022, and lower right panel- Band function. The figure is reproduced from Adhikari et al. [14]. ©AAS. Reproduced with permission

The nature of the line emission versus radius is similar for Mrk 509, NGC 5548, and the Band function, independent of the shape of the illuminating radiation. For all the considered lines, the radial emission profiles display the strong suppression (by a factor of ~ few to a few orders of magnitude) of the line luminosities in a region close to the sublimation radius. The suppression of the line emission is also clearly noticed in total gas emission profile (see the black dashed line in the figure). The importance

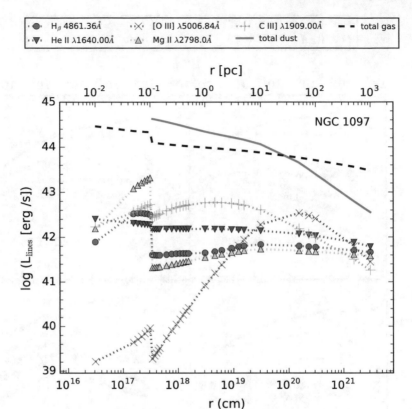

Fig. 5.3 The same as that of an individual panel of Fig. 5.2 but for the spectral radiation shape of LINER NGC 1097

of dust is also displayed on all panels as magenta continuous line. These results are in agreement with the result of Netzer and Laor [1] performed only for one standard AGN SED from Mathews and Ferland [3]. However, the radial luminosity profile of the He II line is only weakly suppressed by the factor of $\sim$3 with the inclusion of dust. This is different from the case of Netzer and Laor [1], where the He II emission goes down by $\sim$ an order of magnitude.

Considerably different radial emission profile for the semi forbidden line C III] and the forbidden line [O III] is obtained in case of NLSy1 PMN J0948+0022 SED. For NLSy1 PMN J0948+0022, the C III] line luminosity decreases steeply with increasing radius showing a different behaviour from the other SEDs, where it increases in a region between $0.1 < r < 10$ pc. As can be seen from the Fig. 5.2 (lower left panel), the [O III] emission is not significant in a region corresponding to the NLR ($r \sim 10 - 100$ pc), which is different from the other SEDs where an excess of [O III] emission is seen at those distances. Interestingly, this result is consistent with the SDSS spectrum of PMN J0948+0022, where [O III] emission is not present ([15], see their Fig. 2, lower panel). Finally, our model shows that He II emission in this object is not significantly affected by the presence of dust. Rather, it is fairly constant close to 0.1 pc and decreases slowly as a function of the radial distance. The differences in line emission of C IV, [O III] and He II from the other SEDs considered in our study are connected with the fact that the PMN J0948+0022 SED has fewer photons in the extreme UV and soft X-ray band. This significantly smaller number of hydrogen ionizing photons causes the drastic drop, by $\sim$2 orders of magnitude, in lines emitted in the lower density dusty medium on the larger radii for this source.

As depicted in the Fig. 5.3, in the NLR regime, the LINER NGC 1097 line emissivities differ from the rest of the SEDs. In this case, the extent at which the C III] emission decreases with radius is intermediate between the cases of PMN J0948+0022 and rest of the SEDs. At distances > 10 pc, the emission of lines He II, Hβ and Mg II remains fairly constant with radii. For the radii >300 pc the He II emission becomes prominent than other lines which is a different behaviour as compared to the other SED cases.

Despite these various differences in the nature of line emission among different sources under consideration, the effect of dust on the line suppression is clearly visible in all cases. The general conclusion that I reached at this stage is—for gas with physical conditions and parameters given by Netzer and Laor [1], the apparent gap in the line emission cannot be removed by changing the shape of the illuminating continuum and there is a clear separation between the BLR and the NLR as observed. The minor differences in the radial luminosity profiles for NLSy1 and LINER SED are not sufficient to explain the origin of ILR.

As a second step in the exploration of line emissivity profiles and the ILR emission, I increased the density at R_d by two orders of magnitude in the power law given in Eq. 5.1. The motivation for increasing the local gas density at R_d is inspired by the recent studies of line emitting gas in some AGNs (see Sect. 1.2 for details). I increased the normalization of the gas density at $r = 0.1$ pc, to be $n_H = 10^{11.5}$ cm^{-3} in our computations. Note that, the ionization parameter changes inversely with n_H as given in Eq. 2.12. All other model parameters are kept the same.

Figures 5.4 and 5.5 depict the simulated line emissivities for the gas density at $n_H = 10^{11.5}$ cm^{-3} at the sublimation radius for five different SED types. The Hβ, He II, and C III] emission lines behave differently from the lower density normalization case. Interestingly, the presence of dust does not suppress C III] and Hβ line emission, rather it is enhanced by the factor of $\sim$1.5 at $0.1 < r < 0.3$ pc, in case of Mrk 509, NGC 5548 and Band function SEDs. For the SED of NLSy1 PMN J0948+0022 and LINER NGC 1097, Hβ emission decreases slightly with the distance, but there is no large jump at the sublimation radius, as was found in the lower density case. Only the Mg II line displays a jump at 0.1 pc, but the jump height is less prominent than in the lower density case. However, in all other SED cases the lack of line suppression suggests the presence of the intermediate line emission.

This result clearly shows that the significant extinction of the line emission caused by the presence of dust beyond the sublimation radius disappears if the gas is highly dense. This clearly indicates that the intermediate line emission is caused by high density gas located at radial distances $r \sim 0.1 - 1$ pc in agreement with recent observations [16–18]. Additionally, this result is in agreement with the speculation

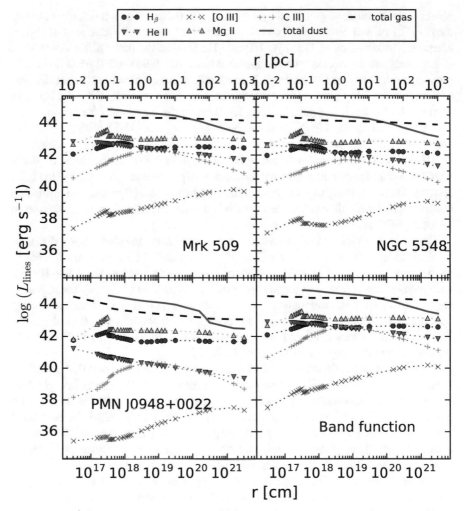

Fig. 5.4 Same as in Fig. 5.2, but for the density $n_H = 10^{11.5}\,\text{cm}^{-3}$ at the sublimation radius, $R_d = 0.1$ pc. The presence of the ILR emission is clearly seen. The figure is reproduced from Adhikari et al. [14]. ©AAS. Reproduced with permission

made by Landt et al. [19], where they discuss the possible role of the high density gas for producing the smooth BLR. Furthermore, if the ILR is located at distances predicted by this model, one can estimate the expected RM lag. Within the framework of this model, the predicted ILR lag would be of the order of 100–1000 light-days.

The physical reason for the density dependence is connected with the thickness of the ionization front of H^+ formed inside the gas of different n_H values, which depends on the ionization parameter U through the Eq. 3.2. The thickness of H^+ front, from which the line emission comes, is larger for the low density gas in comparison with high density gas as shown in the Sect. 3.2. So, the less dense clouds have much larger

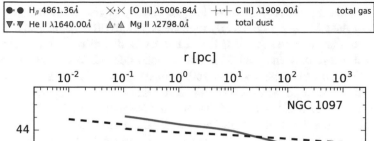

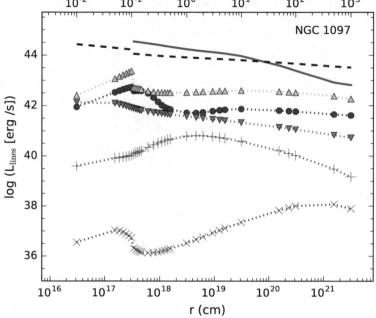

Fig. 5.5 Same as in Fig. 5.3, but for the density $n_H = 10^{11.5}$ cm^{-3} at $R_d = 0.1$ pc

thickness of H$^+$ layer, larger dust column density in the zone with abundant photons, and therefore the dust absorption is significant. This effect is nicely demonstrated in the Fig. 3.2.

Furthermore, Eq. 3.2 can be used for the quantitative estimation of threshold U, below which the gas opacity for incident photons dominates over opacity of dust. Taking recombination coefficient for hydrogen α_B (at 10^4 K) $= 2.6 \times 10^{-13}$ cm^3 s^{-1}, the resulted column density of the H$^+$ layer is: $N_{H^+} \sim 10^{23}$ U. Therefore, the dust optical depth for UV photons is $\tau_{dust} \cong N_{H^+}/10^{21} = 100U$ (for details see Güver and Özel [20]). For $U = 0.1$, $\tau_{dust} = 10$ at the edge of the H$^+$ region. This implies that the width of the H$^+$ layer where gas actively absorbs and emits is only 1/10 of the total layer. In other words, increasing the density by the factor >10 would mean $U < 0.01$ which implies $\tau_{dust} < 1$. This means that the gas opacity always dominates over dust opacities for higher densities and it does not matter if the gas is dusty or not, and therefore no suppression of the line emission is physically possible.

It is worth to mention the conclusion of Ferland and Netzer [21], saying that for the LINERs, if line emission comes from the photoionized gas, the ionization parameter U is $\leq 10^{-3}$. This is less than the threshold U corresponding to higher density ($10^{11.5}$

cm^{-3}) implied by my result, where the gas opacity is always dominant over the opacity of dust. So, my result clearly indicates that LINERs should also exhibit the ILR. Note, that the presence of ILR in LINERs is also discussed by Balmaverde et al. [22] where they analysed 33 LINERs (bona fide AGN) and Seyfert galaxies from optical spectroscopic Palomar survey observed by *HST/STIS*. However, the density of the outer portions of ILR claimed by those authors is about ~3 orders of magnitude less (i.e., $10^4 - 10^5$ cm^{-3}) than what our model predicts. Most probably this is due to the fact that LINERs are much fainter (i.e., $L_{bol} \sim 10^{41}$ erg s^{-1}), 4 orders of magnitude less than the source luminosity considered in this work.

To further investigate the role of gas density in the formation of the ILR, I simulated the line emission for several values of density normalizations: $n_H = 10^{7.0}$, $10^{9.4}$, $10^{10.0}$, $10^{11.0}$ and $10^{11.5}$ cm^{-3}, keeping other parameters the same. Figure 5.6 shows the dependence of line luminosity radial profiles on the density normalization for two representative SEDs: Mrk 509 (left panels) and PMN J0948+0022 (right panels), for all emission lines considered. For the rarefied clouds, the luminosity of some lines decreases many orders of magnitude. Those lines would not be observed due to the sensitivity of the telescopes. Therefore, the lower limit on the line luminosity above which line could be visible, is marked by the horizontal dashed line on each panel. This limit is estimated by assuming a Gaussian spectral profile of the emission lines on top of the underlying continuum given by our SEDs. For this calculation, spectral line width is set to the value corresponding to Keplerian velocity at 1 pc ($\sim ILR$ radii), for a BH mass $\sim 10^8 M_\odot$. It is rather hard lower limit on the observable line luminosities since the simulated isotropic luminosities should be at least 10 times higher (assuming 100% covering factor). Nevertheless, simulated line luminosities presented in Fig. 5.6 fall below this limit in some cases, and those lines have no chance to be detected.

For a given SED, there is a preferred value of density for which the line emission is the highest. For example; the Hβ luminosity increases with the density normalization and is a maximum for $n_H = 10^{11.5}$ cm^{-3} at distances $0.1 < r < 0.4$ pc. The He II luminosity peaks at $r < 0.1$ pc, and the maximum emission occurs for the lower density, whose value depends on the SED shape. In case of PMN J0948+0022 the luminosity peak is lower by an order of magnitude than for Mrk 509 SED, and occurs for density $n_H = 10^7$ cm^{-3}. This is in agreement with Locally Optimised Cloud (LOC) model of Baldwin et al. [23].

The radial distances at which the Hβ and He II emissions are the strongest are in agreement with the results inferred from the RM studies of the BLR in AGN. The radial stratification with ionization potential of the species producing the line has been observed (i.e. Clavel et al. [24], Peterson and Wandel [25]), showing that He II line always originates at distances closer by a factor of three/four to the nucleus than Hβ line.

For the Mrk 509 SED, the emission, with density normalization $n_H = 10^7$ cm^{-3} at 0.1 pc, is insignificant for all considered lines. However, the [O III] emission is the strongest for the PMN J0948+0022 SED with the lowest density which agrees with the prediction that forbidden lines are formed in low density gas. For both SEDs,

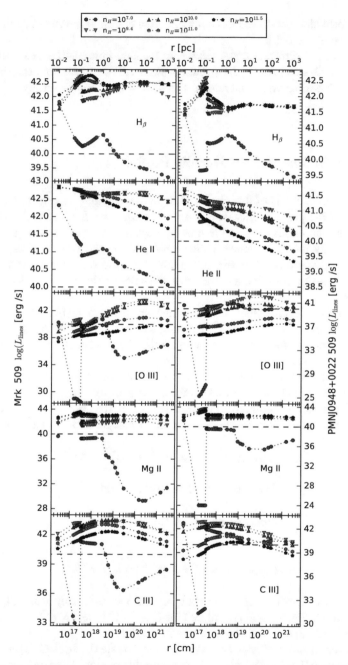

Fig. 5.6 The line emissivity profiles for different density normalizations at R_d: $10^{7.0}$ (red circles), $10^{9.4}$ (green triangles down), $10^{10.0}$ (blue triangles up), $10^{11.0}$ (magenta hexagons) and $10^{11.5}$ (black pentagons). The left and right panels display the case of Mrk 509 and PMN J0948+0022 SED shapes respectively. From top to bottom the emission lines Hβ, He II, [O III], Mg II and CIII] are shown in the order. The horizontal black dashed line shows the detection limit on the luminosity of each line. The figure is reproduced from Adhikari et al. [14]. ©AAS. Reproduced with permission

the semi-forbidden line C III] has a maximum emission at intermediate density i.e,
$n_H = 10^{9.0}$ cm^{-3} at 0.1 pc.

Despite of the different distances where the line luminosities peak, the strong
jump in the emission profiles disappears in almost all cases for both shapes of SED
when density normalization increases. The lack of this jump naturally explains the
origin of the ILR.

5.3   ILR for Power Law Density Profile

In the previous sections of this chapter, I demonstrated that the effect of dust is not
significant if the gas is highly dense. As a consequence, the line emissivity radial
profiles are continuous even if the dust is present in the emission region. In this
section, I study the dependence of line emissivity on the various density profiles
used in the simulations. The main aim at this point is to search how the appearance
of intermediate line emission is sensitive on the value of slope β of the density power
law profile.

Firstly, I consider the density profiles of the type given in the Eq. 5.1 with the value
of normalization $A = 10^{11.5}$ cm^{-3}. I choose this normalization for all simulations
presented in this section after my conclusion that for this density value, the ILR can
physically be present.

For the density profile given by Eq. 5.1, the resulting ionization parameter U
depends on the cloud location and on the amount of ionizing photons (Eq. 2.12).
Assuming the same bolometric luminosity: $L_{bol} = 10^{45}$ erg s^{-1}, in case of four
sources, I obtain the following scaling laws for the ionization parameter with distance
in pc:

$$U_{Mrk509} = 6.72 \times 10^{-6} \, (r/R_d)^{\beta} \, r^{-2}, \tag{5.4}$$

$$U_{NGC5548} = 3.11 \times 10^{-6} \, (r/R_d)^{\beta} \, r^{-2}, \tag{5.5}$$

$$U_{NGC1097} = 1.13 \times 10^{-6} \, (r/R_d)^{\beta} \, r^{-2}, \tag{5.6}$$

and

$$U_{PMNJ0948} = 1.08 \times 10^{-6} \, (r/R_d)^{\beta} \, r^{-2}. \tag{5.7}$$

I adopted the density profiles for six values of $\beta = 0.5, 1.0, 1.5, 2.0, 2.5, 3.0$
as shown in the Fig. 5.7 and considered the four SED types: Mrk 509, NGC 5548,
PMN J0948+0022 and NGC 1097 shown in Fig. 5.1.

In addition to the previously considered lines: Hβ, He II, Mg II, C III] and [O III], I
also included Fe II and C IV in my study onward this section. There is a long standing
problem of thousands of transitions corresponding to the different energy levels of
Fe II emission. Because of these numerous transitions, it is very difficult to identify
the particular line centroid. For our convenience, I consider the wavelength band
λ (4434.00–4684.00) Å and the integrated luminosity in this band is obtained. The

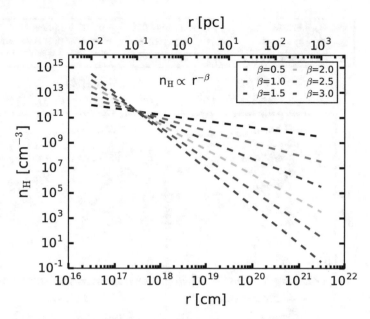

Fig. 5.7 Power law density profiles with varying slopes shown

resulting line emissivity profiles for four cases of SED are presented in the Figs. 5.8, 5.9, 5.10 and 5.11 respectively.

In all cases of density power law slopes, a continuous line emission is recovered, with a small enhancement of the permitted lines Hβ and He II at the radial distance around 0.1 pc corresponding to the intermediate region, independent of the shape of the SEDs in consideration. There is a small reduction of Mg II line at 0.1 pc though not very significant as compared to the suppression presented by Netzer and Laor [1]. However, the Fe II line is always suppressed by the dust. This shows that Fe II emission can be used as a tracer of dust presence in the AGN emission regions. Nevertheless, the semi forbidden line C III] contribution to the intermediate emission component becomes most prominent for the density profile with $\beta = 1.5$.

The most noticeable effects of different density profiles on line emissivities occur in the NLR range, i.e. for $r > 50$ pc. This behavior is quite obvious since for those radii, differences in densities between profiles are the biggest. For $\beta = 0.5$ and 1.0, density falls slowly and remains moderately high across the radii causing the strong suppression of forbidden line [O III]. [O III] is effectively produced in low density environment and becomes prominent when the density around the radius 10 pc becomes low enough, the cases for the profiles with $\beta \geq 2.0$. Narrow line emission is dominated by the C III] components when the density distribution is given by the profiles with $\beta \geq 2.0$. I found that the derived line emissivities for all cases of power law density slopes, particularly in the region of intermediate line emission, do not strongly depend on the shapes of the SEDs used. There are subtle differences in

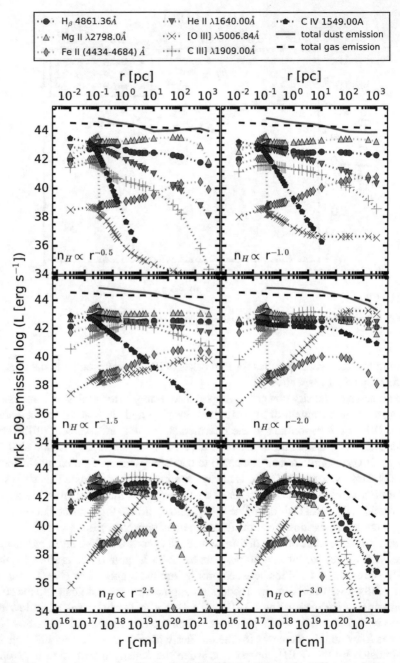

Fig. 5.8 Line emission versus radius for Mrk 509 SED for CD clouds. Different subplots are for various density slopes given at the bottom of each panel. Major emission lines: Hβ (red circles), He II (green triangles down), [O III] (blue crosses), Mg II (cyan triangles up) and C III] (magenta pluses), Fe II (orange diamonds) and C IV (purple pentagons) are shown for each density profile cases. For clarity total dust emission (magenta continuous line) and total gas emission (black dashed line) are also shown. The figure is reproduced from Adhikari et al. [26]

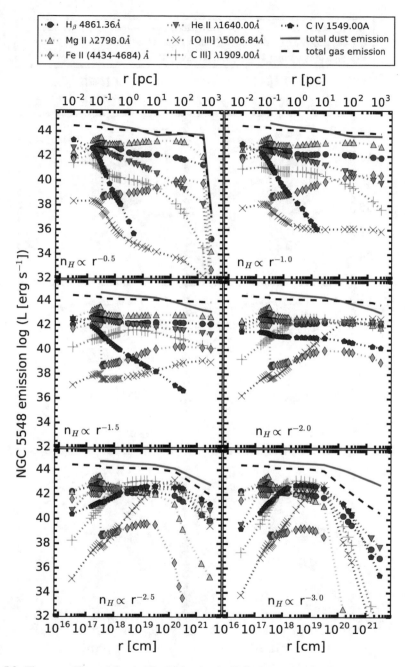

Fig. 5.9 The same CD model as in Fig. 5.8 but for the SED shape of the NGC 5548

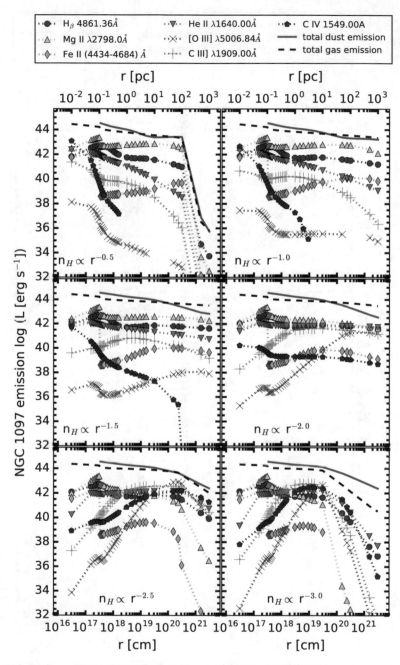

Fig. 5.10 The same CD model as in Fig. 5.8 but for the SED of NGC 1097

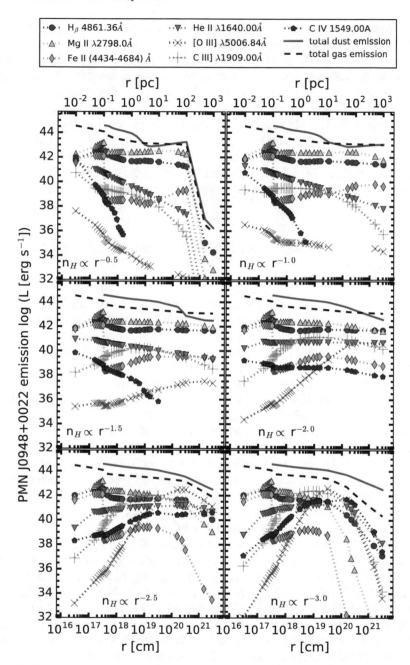

Fig. 5.11 The same CD model as in Fig. 5.8 but for the SED PMN J0948+0022

emissivities corresponding to the BLR and NLR due to the different amount of UV and soft X-ray photons among the SEDs.

This result clearly demonstrates that the different slopes of the density distribution do not change ILR emission significantly as long as the gas density at the sublimation radius is high, in this case being $10^{11.5}$ cm^{-3}. In all cases of considered density profiles, an intermediate line emission around $0.1 - 1$ pc is obtained, mostly manifested in permitted lines Hβ, He II and Mg II, and weakly present in the semi forbidden line C III]. This indicates that the high density and the low ionization environment favors the intermediate line emission rather than the high ionization environment where the forbidden line [O III] is produced. So, in the AGNs where the ILR is seen in observations the emitting region is composed of the dense and less ionized gas.

5.3.1   CP Versus CD for Total Model of Radially Distributed Clouds

In Chap. 4, I have shown that, the observational properties of AGN in the context of ionized absorber can be well reproduced by the assumption that the clouds are in total pressure equilibrium. There are several propositions that the emission gas clouds in pressure equilibrium well explain the observed remarkably uniform line ratios in the Seyfert NLR emission [27–30]. In Sect. 3.3, I demonstrated that the line luminosities emitted from the clouds with CP assumption are higher by $\sim$1.5 order of magnitude than the CD clouds when the less dense gas is considered. This gives a potential probability to distinguish CP clouds from observations. However it is still under discussion whether the emission clouds in AGN are under CP or CD.

In Sect. 3.3, a comparison between the CP and the CD case for a single gas cloud model demonstrated that the requirement for the radiation pressure confinement (RPC) i.e., $P_{rad} >> P_{gas}$ is only a special condition of CP solution, which is usually achieved for the low gas density. However, even if the gas density is high the clouds can still be in CP and differ from the CD case (Fig. 3.3). So it is worth to compare line luminosities derived from CD and CP assumptions.

In this section, I present the radial distribution of CP clouds and compare the resulting line emissivities with the CD cloud models discussed in the Sect. 5.3. For better visibility of the lines behaviour, the LILs: Hβ, Mg II, and Fe II, and the HILs: He II, C III], [O III], and C IV, are organized in two different groups. I limit my presentations only for three representative values of $\beta = 0.5$, 1.5 and 2.5.

Each pair of panel columns in Fig. 5.12 represents the comparison between the model which assumes that each cloud is computed with CD to the model which assumes CP clouds. One can easily see that emission line luminosities do not differ when more physical model of CP is used. Profiles of all emission lines are practically the same, when a comparison is done between the left and right pairs of each column panels. This conclusion is valid for all four spectral shapes used in this thesis as inferred from a comparison between the Figs. 5.9 and 5.13 for NGC 5548, Figs. 5.10

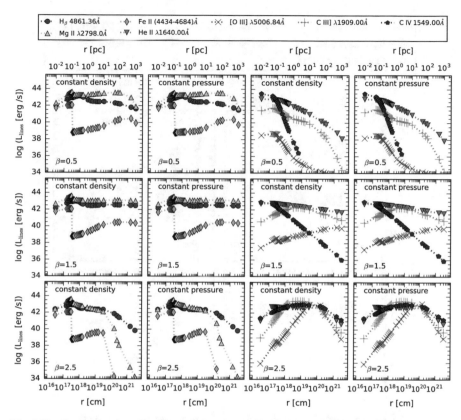

Fig. 5.12 Comparison between the line emissivity profiles obtained with the CD and the CP assumptions in photoionization computations in CLOUDY for the SED shape of Mrk 509. Left two column panels represent LIL: Hβ Mg II and Fe II lines for CD and CP case respectively, while right two columns show HIL: He II, C III, [O III], and C IV lines again for CD and CP respectively. Three row panels show cases for $\beta = 0.5$, 1.5 and 2.5 from top to the bottom. The figure is reproduced from Adhikari et al. [31]. ©AAS. Reproduced with permission

and 5.14 for NGC 1097, and Figs. 5.11 and 5.15 for PMN J0948+0022 respectively. I show that for such cold and dense gas clouds as in the BLR and the NLR, the radiation pressure compression is not that significant as in the case of warm absorbers studied in the Chap. 4.

5.3.2 Dust Sensitive Fe II Line

In all the considered cases in the previous sections, the Fe II line is the only line which shows strong emissivity drop by several orders of magnitude at the sublimation radius. Such behavior predicts the lack of intermediate component for this line. Therefore,

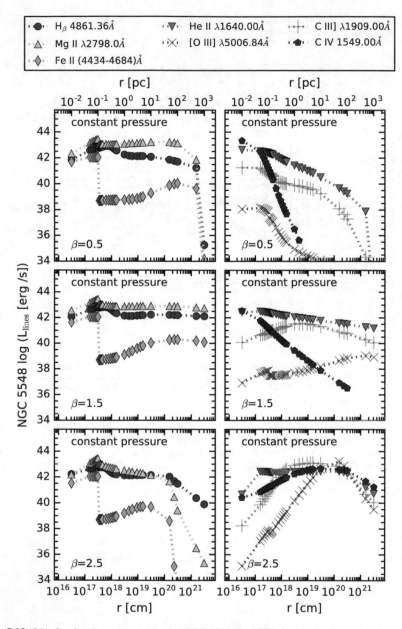

Fig. 5.13 Line luminosity versus radius for NGC 5548, Sy 1 SED for CP clouds. Left column panel represents LIL: Hβ Mg II and Fe II, while right column shows HIL: He II, C III], [O III], and C IV, for constant pressure model. Three row panels show cases for $\beta = 0.5$, 1.5 and 2.5 from top to the bottom. The figure is reproduced from Adhikari et al. [31]. ©AAS. Reproduced with permission

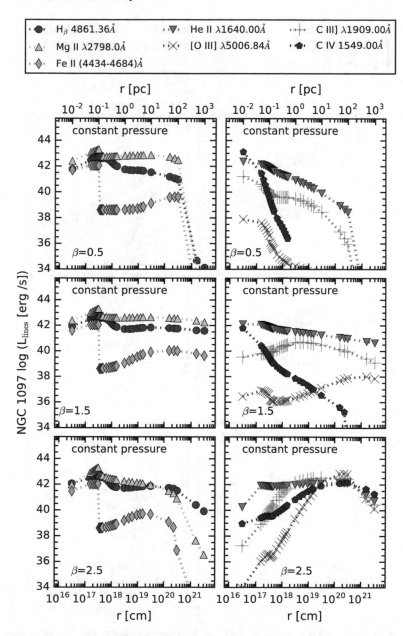

Fig. 5.14 The same CP model as in Fig. 5.13, but for NGC 1097, LINER SED. The figure is reproduced from Adhikari et al. [31]. ©AAS. Reproduced with permission

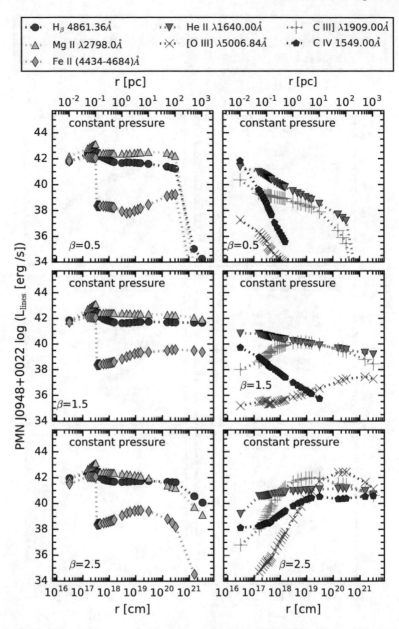

Fig. 5.15 The same CP model as in Fig. 5.13, but for PMN J0948+0022, NLSy1 SED. The figure is reproduced from Adhikari et al. [31]. ©AAS. Reproduced with permission

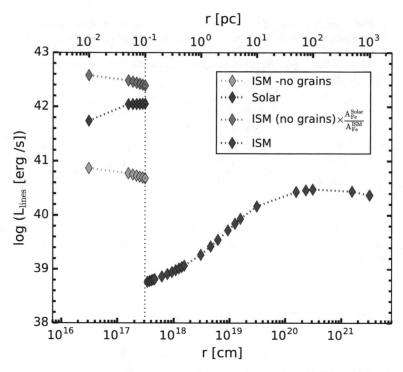

Fig. 5.16 The comparison of Fe II line luminosity for two models of clouds illuminated by Mrk 509 SED with different abundances of dustless clouds. The ISM composition with grains is used for dusty clouds located further than R_d, marked by vertical dotted line. For clouds located closer to SMBH than R_d, I plot the Fe II line luminosity for Solar iron abundance by red diamonds and for ISM without grains abundance by orange diamonds. Green diamonds mark the ISM model without grains multiplied by the Fe abundance ratio of those two models. The figure is reproduced from Adhikari et al. [31]. ©AAS. Reproduced with permission

based on the results of my simulations, the Fe II line is sensitive to the presence of dust and it is not an ILR indicator.

In my model, the strong Fe II emissivity drop may be caused by two effects. The first one is the presence of dust discussed above, and the second effect can be the change of abundances in the gas phase when passing through the sublimation radius. In order to check what really causes the strong drop of Fe II emissivity profile at R_d, I made a test with different abundances in dustless clouds located at BLR. Results are presented in Fig. 5.16, where a comparison of the two models is shown. In the first model, the ISM chemical composition with no grains is assumed for the BLR clouds, while in the second model, the same clouds have typical Solar abundances. In both models ISM composition with grains is used for dusty clouds located further than R_d. It is clearly seen that the presence of dust is responsible for the Fe II emissivity drop

Table 5.1 The major parameters of four AGN used in this thesis. The first and second column lists the name of the AGNs and their types. The integrated bolometric luminosities are given in third column. The masses of the supermassive black hole and the mass accretion rates obtained from the literatures follows in fourth and fifth column respectively. The derived dust sublimation radii using Eq. 2.17 are in column six

Name	AGN type	L_{bol} (erg s^{-1})	M_7^{bh} ($M_\odot$)	$\dot{m}$	$R_{d,17}$ (cm)
Mrk 509	Sy1.5	6.62×10^{45}	14[a]	0.30[b]	31.6
NGC 5548	Sy1	1.28×10^{44}	6.54[c]	0.02[d]	4.41
PMN J0948	NLSy1	2.28×10^{46}	15.4[e]	0.40[f]	58.9
NGC 1097	LINER	9.62×10^{40}	14[g]	0.0064[h]	0.12

[a][35] [b][36] [c][37] [d][18, 38] [e][39] [f][40] [g][41] [h][10]

by about two orders of magnitude, while the change in the iron abundance enhances this effect by one order of magnitude. The conclusion is that Fe II line is a strong dust content indicator in gas emitting clouds in AGN.

5.4 ILR for Disk-Like Density Profiles

In recent years, there have been promising claims that, broad line emission clouds in AGN originate in the winds from the upper part of an accretion disk atmosphere [32, 33]. Moreover, since many years it was expected that the winds from the disk atmosphere exist at different radii from the SMBH [34]. In this section, instead of power law, I consider the disk-like density profile that is expected when the gas clouds are formed above the AD atmosphere.

For the estimation of cloud disk-like density profile, the knowledge of the AD parameters: the mass of the SMBH and the accretion rate is required. I collected these parameters for four AGN types: Sy 1.5 Mrk 509, Sy1 NGC 5548, NLSy1 PMN J0948 and LINER NGC 1097 from the existing observational studies. All of them with relevant references are shown in the Table 5.1. The source luminosity, L_{bol}, given in third column of the Table 5.1 is obtained by the integration of flux in the energy range from $1-10^5$ eV. The dust sublimation radius R_d is computed for each source using the Eq. 2.17 and listed in the 6th column of the Table 5.1. It is clear that the luminosity influences the position of the sublimation radius and it is fully taken into account in further computations. Note that, the column density normalization in Eq. 5.2 is set to $10^{23.4}$ cm^{-2} at R_d for each source.

By using these values of the disk parameters, the vertical structure of the AD atmosphere was solved assuming the standard thin disk model of Shakura and Sunyaev [42] and diffusion approximation of radiative transfer as described in [4]. Those computations take into account the Rosseland mean opacity tables from Alexander et al. [43] which are important for computing the disk density vertical structure since the true absorption opacity can be an order of magnitude larger than the electron absorption opacity [4]. The switch from the radiation pressure dominated regions

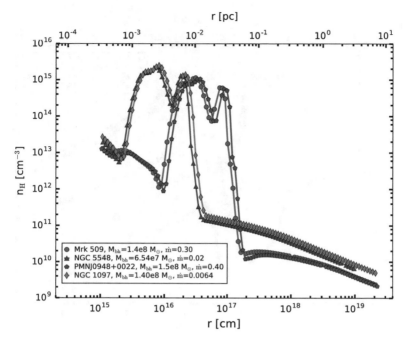

Fig. 5.17 The gas density profiles expected at the optical depth $\tau = 2/3$ in the upper part of the standard AD atmosphere. Parameters used in the computations of the corresponding density profile for each source are also shown. The figure is reproduced from Adhikari et al. [31]. ©AAS. Reproduced with permission

to the gas pressure dominated regions is fully taken into account in the model by assuming that the torque due to viscosity is proportional to the sum of the gas and radiation pressure. However, the model is not valid at larger distances beyond ~1 pc scale as the effects of self gravity as well as the circumstellar cluster's effect becomes significant [44].

The disk-like density profiles at the optical depth $\tau = 2/3$ for all sources are presented in Fig. 5.17. The characteristic feature of such radial density profiles possess strong density peaks up to 10^{15} cm^{-3}, located around the position of BLR, $r \sim 10^{-2}$ pc. This is caused by a strong opacity hump in an accretion disk atmosphere. Outside the density hump its values decline from about 10^{13} cm^{-3} at the distance of 10^{15} cm from SMBH, to less than 10^{10} cm^{-3} at further distances of 10^{19} cm. The corresponding ionization parameter versus radius for each disk-like density profile is presented in the Fig. 5.18. A noticeable difference between ionization degree is seen, which is four orders of magnitude lower for LINER than Sy1 and NLSy1 case. In addition, the density hump is directly reflected in the ionization drop in all sources.

I assume that emitting clouds directly emerge from the disk's atmosphere and preserve its density along the whole range of radii. This is a reasonable assumption at least for low ionization part of the BLR as it may develop as a failed wind embedded in the disk's atmosphere [33]. On the other hand, this assumption is the same as

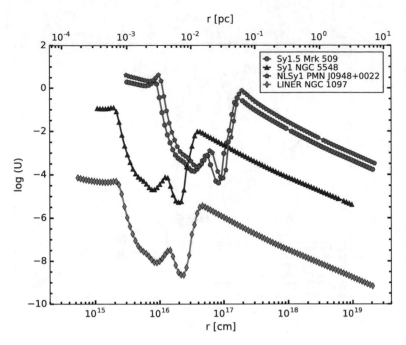

Fig. 5.18 Radial ionization parameter U profiles at the cloud surface computed by Eq. 2.12 for disk-like density profiles as shown in Fig. 5.17. The figure is reproduced from Adhikari et al. [31]. ©AAS. Reproduced with permission

the model of BLR being a part of AD atmosphere [45]. This approach is not in contradiction with line emitting medium geometry similar to bowl on top of the AD [32, 46].

5.4.1 Line Emissivities

Resulting line emissivity profiles for each source: Mrk 509, NGC 5548, PMN J0948+0022 and NGC 1097 are shown in Figs. 5.19, 5.20, 5.21 and 5.22 respectively. Shaded areas in all figures highlight the areas covered by the BLR (pink), ILR (green) and NLR (violet) whose position depends on the AGN types. I assumed that the dynamics of the emitting regions is dominated by the Keplerian motion and computed the corresponding location of those regions assuming the velocity ranges: 15000–3000 km s^{-1} for BLR, 3000–900 km s^{-1} for ILR and 900–300 km s^{-1} for the NLR. In all figures in this section, the radial distance is given in light days for straight comparison with RM studies.

In all AGN types, the emissivity profiles of LILs: Hβ, Mg II and Fe II are insensitive to the density hump in disk-like cloud density profile. On the other hand, HILs:

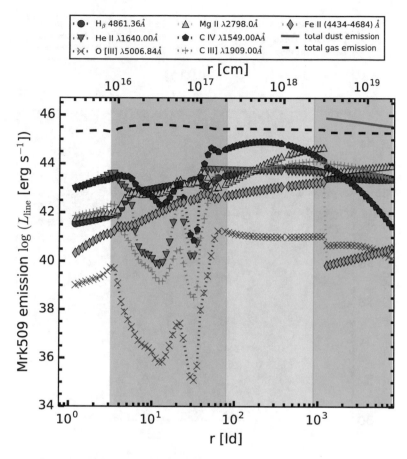

Fig. 5.19 LIL and HIL luminosities versus radius (in light days) obtained for disk-like cloud density profile for Mrk 509. Shaded areas mark the position of BLR, ILR and NLR from left to right respectively, based on the adopted range of Keplerian velocities

He II, C III], [O III], and C IV display strong luminosity drop which reflects the density enhancement in the cloud radial profile. Such HILs luminosity drop is usually situated in the BLR suggesting the division of BLR on two types of low and high ionization as previously suggested by Collin et al. [47], Czerny and Hryniewicz [33].

In case of of Mrk 509 (Fig. 5.19) and PMN J0948 (Fig. 5.21), luminosity is high enough to push the sublimation radius above 0.1 pc making part of the ILR free from dust. This allows for the appearance of emissivity maximum within ILR below the sublimation radius. Thus, in Seyferts, our model predicts dominating intermediate component in LIL: Hβ, Mg II, and Fe II. Their line luminosities rise monotonically upto the sublimation radius. Nevertheless, the emission of HIL: He II, C III], [O III], and C IV is maximum in outer BLR or inner ILR, and decreases with distance.

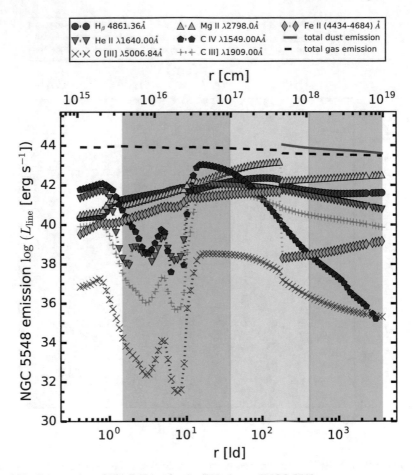

Fig. 5.20 Same as that of Fig. 5.19 but for the SED shape of NGC 5548

In case of LINER shown in Fig. 5.22, dust sublimation radius appears at the position of HIL emissivity drop, caused by cloud density hump because of its low luminosity. The presence of dust in BLR region provides the flat line emissivities all the way to NLR. In those types of sources ILR could be present but not as strong as in other types of AGN.

In general, for disk-like density profile, additional lines as Mg II and slightly Hβ, appear to be dust sensitive. Their luminosity profiles exhibit rapid decrease when dust appears in clouds located relatively far from the center. This does not happen in case of LINER, since dusty clouds are still very dense.

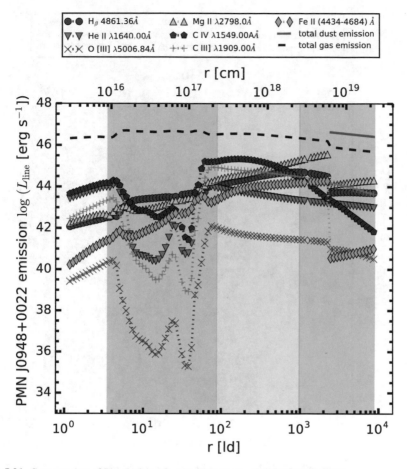

Fig. 5.21   Same as that of Fig. 5.19 but for the SED shape of PMN J0948+0022

5.4.2   Comparison with Observations

Sy1 NGC 5548 is one of the best studied AGN which has been regularly monitored for almost four decades. So the line emission properties deduced for NGC 5548 in this work can be compared with the observational results in more details. An approximate distances of emitting regions from our model can be compared with the results of RM studies. The RM depends on the continuum luminosity measurement [37, 48–50], and data show delay measurements in Hβ span over 6 − 30 days depending on the continuum luminosity. Therefore, in case of Hβ line, I made analysis for continuum luminosity, $L_\lambda(5100$ Å$)$, derived from incident SED (Fig. 5.1), and have checked what radius of the Hβ emission one should expect from observational measurements [51]. For log $L_\lambda(5100$ Å$) \approx 43.25$ used in our model, the delay in Hβ is of the order of 20 days and line width FWHM ≈ 4700 km s^{-1}. Similar results were obtained

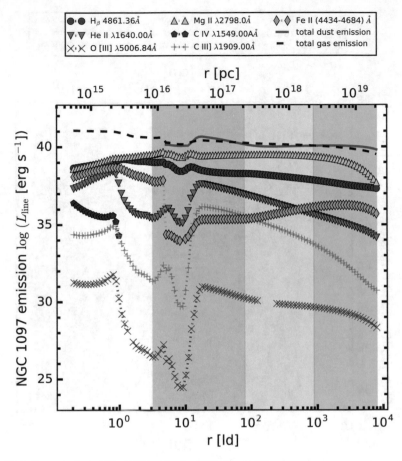

Fig. 5.22 Same as that of Fig. 5.19 but for the SED shape of NGC 1097

by Peterson et al. [52] using ground-based observations made in 1989, where 21 days delay between continuum and Hβ was reported. Those observed parameters correspond to the radius at which the Hβ line luminosity reaches maximum at about 4×10^{17} cm in our model presented in the Fig. 5.20.

However recent observations of velocity resolved RM presented by Peterson et al. [53] show new points which do not agree with the fitted trend of $L_\lambda(5100\,\text{Å})$ versus Hβ delay [51]. Those new points for $\log L_\lambda(5100\,\text{Å}) \gtrsim 43.3$ present Hβ delay of the order of 3 days, so shorter than even for the lowest, $\log L_\lambda(5100\,\text{Å}) \approx 42.5$, luminosity state. Lu et al. [54] derived BLR response delay to the source luminosity change to be 2.4 years. During this time, BLR may be rebuilt under change of radiation pressure, thus making the comparison of our model to the data more difficult.

NGC 5548 emissivity profile of both low ionization lines Hβ and Mg II is very similar in our model (Fig. 5.23 third panel). However it is not exactly the case in RM

Fig. 5.23 LIL luminosities versus Keplerian velocity obtained for disk-like cloud density profiles. Each panel shows one type of AGN in the way that the source luminosity (given in Table 5.1) decreases from top to bottom. Shaded areas mark the position of BLR, ILR and NLR from right to left respectively, based on the adopted range of Keplerian velocities. The figure is reproduced from [31]. ©AAS. Reproduced with permission

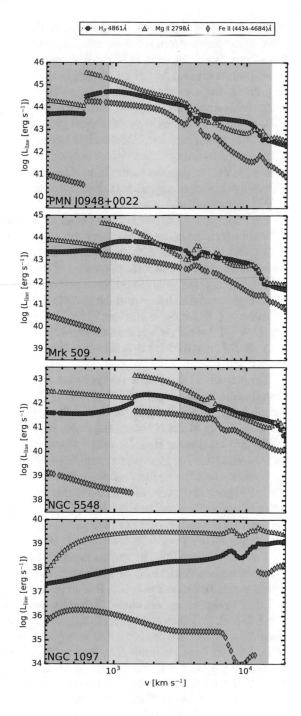

observations. Mg II line is more puzzling in this case. Clavel et al. [24] presented peak-center delays from multi-month IUE campaign done in 1989. Those authors found very broad response in Mg II line covering 34–72 days. In addition, Cackett et al. [55], analyzing Mg II variability have found only weakly correlated broad response to the continuum brightening, with delay response spanning 20–70 days range. Both results may suggest that line luminosity global maximum is located in ILR, which fully agrees with the model presented here. However, even shallower, global line luminosity maximum in Hβ located at the same region of our model is not resolved in the observations of delay. Mg II shows more luminous ILR with maximum before the face of the torus. Brighter ILR in magnesium line should be reflected in higher average delays than in hydrogen Balmer lines and this is the case in RM measurements.

For the case of NGC 5548, Clavel et al. [24] presented light-curves of C III] and C IV for which the delay covers approximately 26–32 days and $8 - 16$ days respectively. This is consistent with the position of global maximum in the emissivities of those lines further on and at the outer edge of dense BLR, which fully agrees with our model. This is also explained by Negrete et al. [56], who inferred emission radius of those high ionization lines from the photoionization condition. They found optimal emission of C IV for $n_H = 10^{12}$ cm^{-3}, log $U \approx -2$ and of C III] for $n_H = 10^{10}$ cm^{-3}, log $U \approx -1.5$ which is consistent with our model as seen in Fig. 5.18. C III] emissivity in ILR is rather flat and allows noticeable emission originating from clouds located at higher radii.

The permitted He II line is always broad and blended with semi-forbidden O III], therefore the measurements of delay covering 4–10 days are more difficult to explain by the models considered in this work. Such delay corresponds to the location of emissivity drop of He II line in the dense BLR in NGC 5548 (see Fig. 5.20). Our model predicts that the He II luminosity local maximum is located on the outer edge of the dense BLR for an expected $\approx$20 days delay.

To check how the position of dusty torus influences radial emissivity profile, in Fig. 5.24 I present the case of two best studied sources computed for 10 times smaller value of sublimation radius given in Eq. 2.18. For Mrk 509, Koshida et al. [57] have reported 120–150 days delay of dust phase, depending on the method of derivation. This value fully agrees with the position of sublimation radius from the formula by Laor and Draine [58], presented in upper panel of Fig. 5.24. However, this fact would eliminate dominance of 1000 km/s component in Mg II line and possibly would remove intermediate width line component of global emissivity maximum of Hβ. This brings our model closer to the observed line delays. In case of NGC 5548, Koshida et al. [57] derived inner torus face radius comparing optical and near infrared variability. Their measurements cover range from 60 to 80 days depending on the NGC 5548 continuum luminosity and method used in calculations. This is over 3 times larger than sublimation radius by Laor and Draine [58] (lower panel of Fig. 5.24), and 2 times smaller than the one computed by Eq. 2.17. The lower position of sublimation radius in NGC 5548 influences maximum emissivity of Mg II shifting it to the lower radii and decrease contribution from 2000 km/s component.

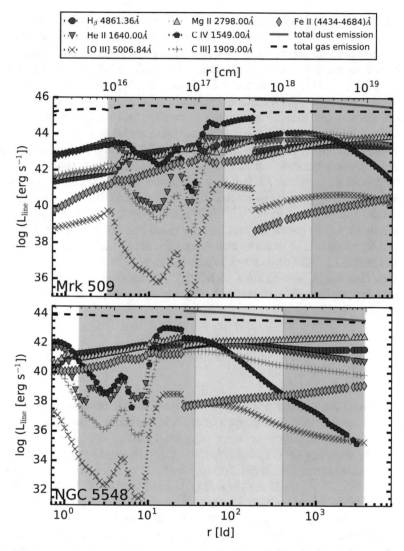

Fig. 5.24 The line luminosity profiles computed using the disk like density profile for two best observed sources, Mrk 509 (upper panel) and MGC 5548 (lower panel). In this case, the dust sublimation radius is computed from the formula in Eq. 2.18. The figure is reproduced from Adhikari et al. [31]. ©AAS. Reproduced with permission

Peterson et al. [59] investigated variability of forbidden [O III] line in NGC 5548. They found delay between 10 and 20 years ($\sim$2–3 pc) and suggested emitting medium with density 10^5 cm^{-3}. While NLR studied in X-rays, as suggested by Detmers et al. [60], covers 1–15 pc or more precisely 14 pc as derived by Whewell et al. [61]. In our model, densities corresponding to the NLR remain high ($\approx 10^9$ cm^{-3}). High emissivities of narrow components in all permitted lines except C IV exist. In addition, [O III] emissivity remains rather low. Crenshaw et al. [62] reported strong narrow components in all optical/UV permitted lines, especially C IV. Thus our model is less accurate reproducing NLR, for assumed disk-like density profile.

Mrk 509 model is in many aspect similar to the NGC 5548. The most noticeable difference is the shift in the emissivity maximum of C III] and C IV toward greater radii. This predicts stronger intermediate emission line components from our model. And this seems to be the case when looking at the observational spectra, presented for example by Negrete et al. [56]. Our model computed for NLSy1–PMN J0948 shows very similar emissivity profile shapes to those of Mrk 509 as those sources have similar BH masses but different SEDs. However, line luminosity to the continuum luminosity ratio is lower for NLSy1, thus effectively broad components blend with continuum and only contrast of narrow components remain sufficient to make the line visible. The LINER case of NGC 1097 is exceptional because the sublimation radius is inside our dense BLR. This fact makes all line luminosity profiles flat up to the NLR, which is in agreement with Balmaverde et al. [22] who emphasized the extended ILR in LINERS up to 10 pc. In case of flat radial luminosity profile, narrow component has the highest contrast, therefore it will dominate line profile, which is in agreement with González-Martín et al. [63], who pointed out that AGN-dominated LINERs are very similar to Seyfert 2 galaxies.

5.5   Conclusions

In this chapter, I investigated the properties of the emission line regions with particular attention to the ILR in different types of AGN by performing the photoionization simulations of the line emitting plasma using the publicly available numerical code CLOUDY. By assuming the radial distribution of the spherical gas clouds, located at different distances from the SMBH, I derived the line luminosities of major emission lines: $H\beta$ $\lambda4861.36$ Å, Fe II $\lambda(4434\text{-}4684)$ Å, He II $\lambda1640.00$ Å, Mg II $\lambda2798.0$ Å, C III] $\lambda1909.00$ Å and [O III] $\lambda5006.84$ Å produced due to the photoionization caused by the radiation energy emitted from the AGN center.

The influence of the model parameters on the line emission is studied by (i) considering the variation in SED shapes by taking different types of AGN: Sy1.5 Mrk 509, Sy1 NGC 5548, NLSy1 PMN J0948+0022, Band SED and LINER NGC 1097, (ii) changing the density normalizations at the dust sublimation radius, (iii) performing computations for CP and CD cloud model, (iv) searching the influence of power-law density slope on the total line emission, (v) considering the disk-like cloud density profile from AD atmosphere, and (vi) using self-consistent source luminosities and therefore the position of the sublimation radius. The comparison of the line luminosities derived from the computed models with that available from the published literatures is done to reach the following conclusions:

1. The presence or absence of the ILR is not determined by the spectral shape of the incident continuum. Different SEDs do produce considerably different behaviour

of the emission line radial profiles, due to the different amount of extreme UV and soft X-ray photons in its broad band SED. However these differences do not adequately explain why an intermediate line emission is observed in some objects.

2. With higher densities normalizations, i.e., $n_H \geq 10^{11.5}$ cm^{-3} at $r = 0.1$ pc, the flat luminosity line radial profiles are obtained for almost all lines and SEDs. Thus, the dust does not suppress the line emission, contrary to the result obtained by [1]. Therefore, the existence of ILR in some source can be explained. When the density of the gas is high enough, emission lines of intermediate velocity width can be produced. Such ILR is predicted to be located at the radial distances $r \sim 0.1 - 1$ pc, and the expected RM lag would be of the order of 100–1000 light-days.

3. I demonstrated the dependence of line luminosity profiles on the density normalization. It is found that the significant line emission in objects with a particular SED occurs at different densities in agreement with RM studies of Hβ, Mg II, [O III] and He II lines.

4. In case of clouds located at distances considered in this thesis, CD and CP cloud models reproduce exactly the same line luminosity profiles in the regime of lines observed in optical/UV. It is caused by the fact that our clouds are dense, but lower density clouds do not produce the ILR.

5. The varying slope of the power law density profile does not affect the nature of the ILR. In particular, the intermediate emission in Hβ is present for all the slopes independent of the SED shape.

6. For the lower slope of the density profile, forbidden [O III] and semi-forbidden C III] lines are strongly suppressed because of the high density environment. As the slope becomes more steeper, i.e. density strongly decreases, these lines become prominent at radial distances corresponding to NLR.

7. Fe II emission line appeared to be most sensitive to the dust presence, since its luminosity drops by two orders of magnitude at the sublimation radius.

8. The density hump in the disk-like density profiles causes mild enhancement of Mg II, Hβ and Fe II lines, while He II, C III] and [O III] are suppressed at the density hump location. This result is consistent with separation of LIL and HIL clouds in two-component BLR model [64].

9. The low luminosity of the LINER NGC 1097 shifts the dust sublimation radius toward smaller distances from SMBH, which makes the emissivity profiles of all lines flat. Therefore, intermediate line component can be detectable, but is less prominent than the narrow line component.

10. The distance inferred from the time delay of Hβ, Mg II, in NGC 5548 taken from RM closely agrees with the distance at which the Hβ line peaks in the simulated line emissivity profile for this source.

11. The NLR from our disk-like model is denser as it is postulated from observations. NLR clouds may become rare while escaping from AD atmosphere, which requires more sophisticated models than I have used in this thesis.

References

1. Netzer H, Laor A (1993) ApJ 404:L51
2. Rees MJ, Netzer H, Ferland GJ (1989) ApJ 347:640
3. Mathews WG, Ferland GJ (1987) ApJ 323:456
4. Różańska A (1999) MNRAS 308:751
5. Nenkova M, Sirocky MM, Nikutta R, Ivezić Ž, Elitzur M (2008) ApJ 685:160
6. Band D et al (1993) ApJ 413:281
7. Kaastra JS et al (2011) Aap, 534, A36
8. Mehdipour M et al (2015) A&A 575:A22
9. D'Ammando F et al (2015) MNRAS 446:2456
10. Nemmen RS, Storchi-Bergmann T, Eracleous M (2014) MNRAS 438:2804
11. Ho LC (1999) ApJ 516:672
12. Nemmen RS, Storchi-Bergmann T, Yuan F, Eracleous M, Terashima Y, Wilson AS (2006) ApJ 643:652
13. Wu Q, Yuan F, Cao X (2007) ApJ 669:96
14. Adhikari TP, Różańska A, Sobolewska M, Czerny B (2015) ApJ 815:83
15. Tanaka M et al (2014) ApJ 793:L26
16. Mason KO, Puchnarewicz EM, Jones LR (1996) MNRAS 283:L26
17. Crenshaw DM, Kraemer SB (2007) ApJ 659:250
18. Crenshaw DM, Kraemer SB, Schmitt HR, Kaastra JS, Arav N, Gabel JR, Korista KT (2009) ApJ 698:281
19. Landt H, Ward MJ, Elvis M, Karovska M (2014) MNRAS 439:1051
20. Güver T, Özel F (2009) MNRAS 400:2050
21. Ferland GJ, Netzer H (1983) ApJ 264:105
22. Balmaverde B, Capetti A, Moisio D, Baldi RD, Marconi A (2016) A&A 586:A48
23. Baldwin J, Ferland G, Korista K, Verner D (1995) ApJ 455:L119
24. Clavel J et al (1991) ApJ 366:64
25. Peterson BM, Wandel A (1999) ApJ 521:L95
26. Adhikari TP, Różańska A, Hryniewicz K, Czerny B, Ferland GJ (2017) Front Astron Space Sci 4:19
27. Dopita MA, Groves BA, Sutherland RS, Binette L, Cecil G (2002) ApJ 572:753
28. Groves BA, Dopita MA, Sutherland RS (2004a) ApJS 153:9
29. Groves BA, Dopita MA, Sutherland RS (2004b) ApJS 153:75
30. Baskin A, Laor A, Stern J (2014) MNRAS 438:604
31. Adhikari TP, Hryniewicz K, Różańska A, Czerny B, Ferland GJ (2018a) ApJ 856:78
32. Gaskell CM (2009) New A Rev 53:140
33. Czerny B, Hryniewicz K (2011) A&A 525:L8
34. Elvis M (2004) In: Richards GT, Hall PB (eds) Astronomical society of the pacific conference series, vol 311, AGN Physics with the Sloan Digital Sky Survey, p 109
35. Mehdipour M et al (2011) A&A 534:A39
36. Boissay R et al (2014) A&A 567:A44
37. Bentz MC et al (2007) ApJ 662:205
38. Ho LC, Kim M (2014) ApJ 789:17
39. Foschini L et al (2011) MNRAS 413:1671
40. Abdo AA et al (2009) ApJ 699:976
41. Onishi K, Iguchi S, Sheth K, Kohno K (2015) ApJ 806:39
42. Shakura NI, Sunyaev RA (1973) A&A 24:337
43. Alexander DR, Rypma RL, Johnson HR (1983) ApJ 272:773
44. Thompson TA, Quataert E, Murray N (2005) ApJ 630:167
45. Baskin A, Laor A (2018) MNRAS 474:1970
46. Goad MR, Korista KT, Ruff AJ (2012) MNRAS 426:3086
47. Collin S, Kawaguchi T, Peterson BM, Vestergaard M (2006) A&A 456:75
48. Peterson BM et al (2004) ApJ 613:682

49. Bentz MC, Peterson BM, Pogge RW, Vestergaard M, Onken CA (2006) ApJ 644:133
50. Denney KD et al (2009) ApJ 704:L80
51. Kilerci Eser E, Vestergaard M, Peterson BM, Denney KD, Bentz MC (2015) ApJ 801:8
52. Peterson BM et al (1991) ApJ 368:119
53. Pei L et al (2017) ApJ 837:131
54. Lu K-X et al (2016) ApJ 827:118
55. Cackett EM, Gültekin K, Bentz MC, Fausnaugh MM, Peterson BM, Troyer J, Vestergaard M (2015) ApJ 810:86
56. Negrete CA, Dultzin D, Marziani P, Sulentic JW (2013) ApJ 771:31
57. Koshida S et al (2014) ApJ 788:159
58. Laor A, Draine BT (1993) ApJ 402:441
59. Peterson BM et al (2013) ApJ 779:109
60. Detmers RG, Kaastra JS, McHardy IM (2009) A&A 504:409
61. Whewell M et al (2015) A&A 581:A79
62. Crenshaw DM, Boggess A, Wu C-C (1993) ApJ 416:L67
63. González-Martín O et al (2015) A&A 578:A74
64. Collin-Souffrin S, Dyson JE, McDowell JC, Perry JJ (1988) MNRAS 232:539

Chapter 6
Summary and Future Studies

Abstract This Chapter contains the summary of the thesis and the possible future extensions of this work.

The most likely scenario of the absorption and emission lines production in the AGN environment is due to the photoionization process caused by the interaction of the AGN continuum radiation with the materials present there. In my thesis, I studied the properties of the warm absorbing medium and the line emitting medium in AGNs where the absorption lines and the emission lines are produced. I utilized the numerical codes CLOUDY and TITAN to simulate the photoionization process and computed the line emissivity profiles as well as absorption measure distributions, and the obtained outputs are compared with the results from observations.

In Chap. 2, I presented the basic equations that are used in solving the photoionization process in both codes CLOUDY and TITAN. I described the parameters required to set up the simulations and introduced the quantities that can be derived and compared with the observations.

In Chap. 3, I discussed the importance of the gas density parameter used in the photoionization computations. I showed that the depth of the H-ionization front is smaller for the high density gas as compared to the less dense gas. This dependence affects the line emission properties since line emission comes from the ionized part of the gas.

I showed that the RPC is only a special case of CP solution, with assumption that $P_{rad} >> P_{gas}$. For the given SED, the RPC condition depend on gas density used in the model computations. For the Mrk 509 like SED, the heating-cooling mechanism significantly depends on the gas density. Deep inside the dense gas clouds, the free-free process becomes the significant gas heating-cooling mechanism and the ionization structure depends explicitly on the gas density. For the parameter space satisfying RPC condition, I showed that TITAN treats thermal instabilities more accurately than the CLOUDY because of the better numerical scheme ALI used in TITAN for radiative transfer computations.

In Chap. 4, I studied the AMD of Mrk 509 using the numerical code TITAN. I showed that the AMD depends on the density. The best AMD model for Mrk 509 is found for the gas density 10^8 cm^{-3} which agrees well with the observed AMD

© Springer Nature Switzerland AG 2019
T. P. Adhikari, *Photoionization Modelling as a Density Diagnostic
of Line Emitting/Absorbing Regions in Active Galactic Nuclei*,
Springer Theses, https://doi.org/10.1007/978-3-030-22737-1_6

obtained by Detmers et al. [1], from the point of the location of dips and their depth. Nevertheless, the modelled AMD for this source has 30 times higher normalization than the observed AMD.

I studied the influence of the SED shapes on the AMD nature by taking significantly different radiation types: SED A and SED B which differ by the ratio of optical/UV flux to the X-ray flux. I found that the absorbing cloud under constant pressure, on illumination with SED A with $L_{disk}/L_X = 100$ produces the average AMD normalization in the range $10^{21} - 10^{22}$ providing an excellent agreement with the observationally obtained value for 6 Sy1 galaxies by Behar [2]. Moreover, due to the softness of the SED A, I found that AMD models depend on the gas density n_H used in the computations. This dependence is related to the fact that when the SED is dominated by soft photons, there is an explicit dependence of heating-cooling mechanism on the gas density, as shown previously by Różańska et al. [3], Chakravorty et al. [4]. For the gas density 10^{12} cm^{-3}, I found the AMD model that matches to the AMD shape obtained for the 6 Sy1 galaxies by Behar [2] up to the ionization parameter $\xi = 4.5$.

I have estimated the wind location with the use of AMD. For the best AMD model in case of Mrk 509, the obtained location of WA is at $r = 1.39 \times 10^{16}$ cm, about 10–100 times closer to the nucleus than the upper limit for the wind location found by Kaastra et al. [5] for 5 ionization components from variability method. My result agrees with the position of BLR derived by RM technique which is of the order of 10^{16} cm. Nevertheless, in case of dense warm absorber, which fits observed AMD for 6 Sy1 galaxies, the wind location obtained in this thesis is $r \sim 10^{15}$ cm which is about 30 gravitational radii for a black hole of 10^8 M$_\odot$. Such close wind location needs further consideration, especially stability analysis.

I devoted Chap. 5 to study the line emissivity radial profiles in different AGN types with particular emphasis on the ILR emission. I considered the major emission lines: Hβ $\lambda4861.36$ Å, He II $\lambda1640.00$ Å, Fe IIλ (4434.00–4684.00) Å, Mg II $\lambda2798.0$ Å, C IV $\lambda1549.00$ Å, C III] $\lambda1909.00$ Å and [O III] $\lambda5006.84$ Å in different types of AGN and computed the line luminosity at various radii from BLR out to the NLR using the numerical code CLOUDY. I demonstrated that the ILR emission seen in the observed spectra of some AGN is not due to the difference in the SED shapes but it is related to the gas density. My calculations showed that the assumption of high density gas in the emission region gives rise to the continuous line emissivity radial profiles in all the considered AGN types. For the gas clouds with density $n_H \geq 10^{11.5}$ cm^{-3} at the dust sublimation radius, the presence of intermediate line emission is seen in the line emissivity radial profiles derived from the photoionization simulations. This is possible as the effect of dust extinction becomes insignificant for the high gas density.

The variation of density prescriptions used in the photoionization computations of the line emitting regions demonstrated that the ILR emission does not depend on the type of the density profiles used. This is valid as long as I assume the high density gas at the dust sublimation radius. For these density profiles used, the results are not sensitive to the different assumptions CP and CD under which the gas operates. This is due to the fact that the computations made here are still in the regime where gas pressure dominates over the radiation pressure. My research showed that Fe II emission line is highly sensitive to the dust and can be used as a potential tracer of the dust in AGN environment.

I considered the disk-like gas density profiles expected in the accretion disk atmosphere computed using the realistic parameters derived from observations for each sources and incorporated them in the photoionization computations of the radially distributed emission clouds. The derived line luminosity radial profiles for each source support for the two component BLR of Collin-Sourin et al. [6] where the LIL and HIL clouds are seperated. I found in case of LINER, that the dust sublimation radius is shifted towards SMBH and makes the line emissivity profile flat for all lines. This shows that the ILR component in LINER is detectable but less prominent than the NLR component. The distance inferred from the time delay of $H\beta$, Mg II, in NGC 5548 taken from RM closely agrees with the distance at which the $H\beta$ line peaks in the simulated line emissivity radial profile derived in this thesis.

6.1 Future Studies

The results presented in this thesis open the new possibilities for the future extensions of this work in several directions; both the theoretical and observational.

1. I have shown that the ratio of FeII/$H\beta$ lines is higher for CP models than for CD models for $n_H = 10^5$ cm^{-3}. This result should be checked for higher densities of BLR, when clouds are dominated by radiation pressure. This may potentially explain the observed value of the above ratio which is always too high from the one predicted by CD models.
2. In order to further investigate the AMD as a potential density diagnostic, it is necessary to derive AMDs for many AGNs and make a systematic study of its nature, in particular the universality of normalizations and the number of dips present in the distribution.
3. While deriving the AMD from observations, it is worth to consider using the optically thick clouds in fitting the individual absorption lines. This will properly take into account the fact that the observed absorption lines may be saturated. The ionic column densities derived in this way may influence the AMD normalization.
4. One limitation of the current TITAN code is, it cannot compute the structure of the gas clouds below the limiting temperature 8000 K. This limitation puts the lower limit of $\xi \sim 4$, below which the ionization and thermal structure cannot be computed. However, the observed AMD spans to much lower ionization values i.e., upto $\xi \sim 0.1$. The modification of the TITAN code is needed for valid computations of optically thick medium with temperature below 8000 K.
5. To fully explore thermal instability, the new numerical scheme for simulations of radiative transfer through the multi-phase medium should be developed, which is a long standing project. In particular, the accurate radiative transfer from TITAN code should be combined with extremely rich atomic database from CLOUDY code.

Furthermore, the simulation should be made with adaptive mesh refinement grid in the gas structure, since all ionization fronts are very steep.

6. From observations, it is not yet clear how does the gas density vary radially from BLR through the NLR. As shown in this thesis, the assumption that the emission line regions are spatially originated from the upper part of an accretion disk atmosphere produces the line emissivities fairly consistent with the observations. Nevertheless, our model does not include the screening of continuum radiation by the previous gas clouds located closer to the source. I plan to improve the current model to account for the radiation screening in the future studies.

7. In the models of emission line regions presented in this thesis, all the dust components are assumed to evaporate at the same sublimation radius. However, the dust evaporation depends on the size and types of the grains and the sublimation radius might be at different positions for different grain types. I plan to take this into account in the future work.

8. I have shown that Fe II line is very sensitive to the dust content, and therefore it can be used as a potential dust diagnostic in AGN. Therefore, observational studies should be done in this direction with the aim to determine dust sublimation radius in those sources.

9. In this work, all the gas clouds were assumed to be static. However, in reality the emission gas clouds in AGN are always in motion. In the future work, the dynamics of the gas clouds should be included.

References

1. Detmers RG et al (2011) A&A 534:A38
2. Behar E (2009) ApJ 703:1346
3. Różańska A, Kowalska I, Gonçalves AC (2008) A&A 487:895
4. Chakravorty S, Kembhavi AK, Elvis M, Ferland G (2009) MNRAS 393:83
5. Kaastra JS et al (2012) A&A 539:A117
6. Collin-Souffrin S, Dyson JE, McDowell JC, Perry JJ (1988) MNRAS 232:539

Printed in the United States
By Bookmasters